Geographic Information Systems Science

Geographic Information Systems Science

Editors

Pooja Rajput and Atul Singh

scitus
academics

Geographic Information Systems Science

Edited by **Pooja Rajput and Atul Singh**

ISBN: 978-1-68117-140-1

Library of Congress Control Number: 2015936560

SCITUS Academics LLC,
www.scitusacademics.com
Box No. 4766, 616 Corporate Way,
Suite 2, Valley Cottage,
NY 10989

Printed in United States of America on Acid Free Paper ∞

Contents

Preface

A geographic information system (GIS) is a system designed to capture, store, manipulate, analyze, manage, and present all types of spatial or geographical data. The acronym GIS is sometimes used for geographical information science or geospatial information studies to refer to the academic discipline or career of working with geographic information systems and is a large domain within the broader academic discipline of Geoinformatics. What goes beyond a GIS is a spatial data infrastructure, a concept that has no such restrictive boundaries. In a general sense, the term describes any information system that integrates, stores, edits, analyzes, shares, and displays geographic information. GIS applications are tools that allow users to create interactive queries (user-created searches), analyze spatial information, edit data in maps, and present the results of all these operations. Geographic information science is the science underlying geographic concepts, applications, and systems.

Editor

Chapter 1

Harnessing Earth Observation and Satellite Information for Monitoring Desertification, Drought and Agricultural Activities in Developing Countries

Humberto Barbosa[1], Carolien Tote[2], Lakshmi Kumar[3], and Yazidhi Bamutaze[4]

[1]Universidade Federal de Alagoas (UFAL), LAPIS, Brazil

[2]Flemish Institute for Technological Research (VITO), Centre for Remote Sensing and Earth Observation, Boeretang, Mol, Belgium

[3]Atmospheric Science Research Laboratory, SRM University, India

[4]Department of Geography, Geo-Informatics and Climatic Sciences, Makerere University, Kampala, Uganda

INTRODUCTION

With the drastic advances in technology over the past decades, the availability, as well as the quantity, of large data sets for research in almost every scientific field has increased dramatically. More specifically, the availability of earth observation-based imagery data and satellite information for research purposes and practical applications has grown with many organizations such as the European Organisation for the Exploitation of Meteorological Satellites (EUMETSAT), the National Aeronautics Space Administration (NASA), the National Oceanic and Atmospheric administration (NOAA), the Flemish Institute for Technological Research (VITO), etc. for example, through GEONETCast, which is part of the core Global Earth Observation System of Systems (GEOSS), the users do not need to repeatedly build ground receiving stations for different satellites [1]. However, despite a wealth of remotely sensed data provided by GEONETCast, investments in science technology and innovations is often a low priority for decision and policy makers in most developing countries in Africa, Asia and global emerging economies like India and Brazil. Yet, most of these countries face serious environmental risks and development challenges which require reliable and timely access to accurate Earth Observation (EO) data and derived environmental information for their sustainable development. In particular, there is a clear need for research on the integration and utility of remote sensing data and products into the risk assessment cycle, scenario development and impact forecasting, in view of global (climate) change [2].

Climate alterations, although global in nature, may have different impacts in different regions of the world. Reports of Intergovernmental Panel on Climate Change (IPCC) showed that the global average surface temperature has increased over the 20th century by about 0.6°C [3]. Global Circulation Models have projected that this rise in temperature may increase to a range of 1.5 – 5.8°C by the end of the century. Results of crop growth modeling under climate change scenarios suggest that agriculture, and thus human well-being, will be negatively affected by climate change, especially in developing countries [4]. The vulnerability to drought and land degradation has increased in the past decades and this is especially true due to increased population pressure and limited livelihood options in drought-prone areas [5]. Significant gaps in observing systems exist, especially in developing countries, and timely access to both surface-based and space observations is still a challenge in many locations.

Global-scale population growth and economic development will have a large impact on water supply and demand, and it is necessary to understand the interactions between climate change and variability, hydrology and human systems, in order to have a view on future water vulnerabilities [6]. Since developing countries will become more susceptible to climate variability and drought, it is essential to develop climate (impact) monitoring services. A climate service involves broad partnerships of producer and user organizations, climate scientists, climate service providers, economists and social scientists. It provides an opportunity to interlink global, national and regional information systems; to provide essential information to policy makers, decision takers and to the public in general at regional and local scales, and a provide for a distributed decision-relevant research and development capability [7]. The climate service for the developing countries might focus on collaborative problem solving. Also capacity building and the improvement of infrastructure, related to the acquisition of advanced remote sensing technologies and the installation of satellite receiving stations by, is needed.

The gist of this chapter s therefore to strengthen the capacities of the regional scientific community to provide stakeholders from drought to agricultural activities with satellite information that is directly useful to improve decision-making in the context of the developing countries. The chapter is based on four case studies from Africa, South America, India and Europe to demonstrate the utility of satellite imagery data obtained from free or low cost platforms in providing information to address the above environmental issues, which are critical particularly in developing countries. Based on the expertise, experience and interest of a regional network of scientists from Uganda, Brazil, India, and Europe, the chapter focuses on harnessing satellite remote sensing resources and products for drought monitoring of areas subject to or in risk of land degradation processes, agricultural productivity and drought assessment. The following paragraphs include a series of example cases where time series of satellite imagery is used to monitor the impact of climate (change):

- Degradation monitoring over South America

- Relation of NDVI with the moisture Index over the regions of different climatic types
- Sugarcane yield estimation and modelling based on NDVI data over Southeastern Brazil and
- Production and yield estimates of five major crops over Uganda

The importance of each of the above aspects in the context of assessing, monitoring and managing the natural hazards such as desertification, drought and risk management are discussed in detail in the individual sections.

All case studies make use of time series of Normalized Difference Vegetation index (NDVI), an index of vegetation activity that can be derived from broad band measurements in the visible and infrared channels onboard satellite instruments and which is directly related to the photosynthetic capacity of plants [8]., Satellite sensors such as NOAA AdvancedVery High Resolution Radiometer (AVHRR), Moderate Imaging Spectroradiometer (MODIS), Système Pour l'Observation de la Terre (SPOT) Vegetation etc. provide NDVI data on different intervals (8-day, 10-day, monthly and seasonal). NDVI value varies minus one (-1) to plus one (+1), whereby low NDVI values (< 0.2) reflect sparse vegetation, and higher NDVI values (> 0.4) reflect high vegetation densities. Several studies looked into the interannual variability and trends of NDVI in relation to meteorological parameters such as rainfall, temperature etc [9-11].The long time series of NOAA AVHRR has been widely used to relate the synoptic meteorology/climatology to understand the vegetation dynamics, vegetation response to climate and climate vegetation feedback mechanism [5, 12-14]. The studies of [9, 15-17] concluded that AVHRR NDVI is a valuable tool to monitor and asses large scale agricultural droughts. Other studies use NDVI time series derived from MODIS [18-20] or SPOT-Vegetation [21, 22].

In this work, the authors made use of NDVI imagery datasets derived from different satellites to relate the respective crop and climatic parameters to understand the sensitivity of NDVI with these parameters and further to use them for vegetation monitoring over large areas which contributes to risk management.

TIME SERIES ANALYSIS FOR DESERTIFICATION MONITORING

In the last decades, remote sensing technologies started to contribute enormously in documenting changes in land cover and monitoring desertification, drought and agricultural activities on regional and global spatial scales. Although desertification is highlighted as one of the most important global environmental issues, both desertification and greening processes have been reported on global scale [23-26]. These processes are related in many ways with other environmental issues, such as climate change and the carbon cycle, loss of biodiversity and sustainability of agriculture [27]. Also in South America environmental change is an important concern. In the last decades, South American ecosystems underwent important functional modifications due to climate alterations and direct human interventions on land use and land cover [28]. In South America, the main forest conversion process in the humid tropics in the period 1990-1997 was the clear-cutting of closed, open, or fragmented forest to make room for agriculture at a rate of approximately 1.7 million ha per year [29]. Apart from deforestation, also forest degradation occurs, a process leading to a temporary or permanent deterioration in the density or structure of vegetation cover or its species composition. Land degradation in arid, semi-arid and dry sub-humid areas is called desertification, and is the result of various factors, including climatic variations and human activities [30].

To determine desertification conditions, this case study focuses on vegetation dynamics in South America over a long time period based on a time series of low spatial resolution, high temporal resolution NDVI derived from SPOT-Vegetation, and to recognize to which extent this variability can be attributed to variability in rainfall, since rainfall is one of the most determinant factors of vegetation growth. In general, and especially in semi-arid regions, strong correlations between precipitation and the NDVI can be found. Therefore, the NDVI can be used as an indicator for vegetation status and vegetation response to precipitation variability. This study expands the analysis, as was performed on the Andes region [31] to South American continental level. In the first phase, trends of vegetation and precipitation indices are analyzed. In a next step, through correlation of NDVI and precipitation dynamics, these areas where the evolution of vegetation is not related to climate only and human induced impacts play an imperative role can be identified.

The time series of SPOT-Vegetation 10-daily composite NDVI data (April 1998 – March 2012) at 1 km resolution (http://www.vgt.vito.be/) was smoothed [32] and consequently synthesized to monthly images using the maximum value composite technique. Also 10-daily rainfall estimates at 0.25° resolution, available from the European Centre for Medium-Range Weather Forecasts (ECMWF) through MeteoConsult and the Monitoring Agricultural ResourceS (MARS) unit, were combined to retrieve monthly composites. The spatial resolution of the NDVI time series was degraded in order to fit the rainfall estimates using a weighted average approach. Many authors remove seasonality by integrating the data into annual values (e.g. [25, 28, 33]). In this study, in order to remove seasonal vegetation changes and thus facilitate the interpretation through the historical record, deviations from the 'average' situation were calculated for the NDVI time series using the Standardized Difference Vegetation Index (SDVI) [34] and for the precipitation time series using the Standardized Precipitation Index (SPI) [35]. In a next step, for each pixel a correlation analysis is performed on the monthly NDVI and SPI datasets, in order to identify the temporal scale at which the environment is most sensitive to precipitation anomalies (the so-called 'best lag').

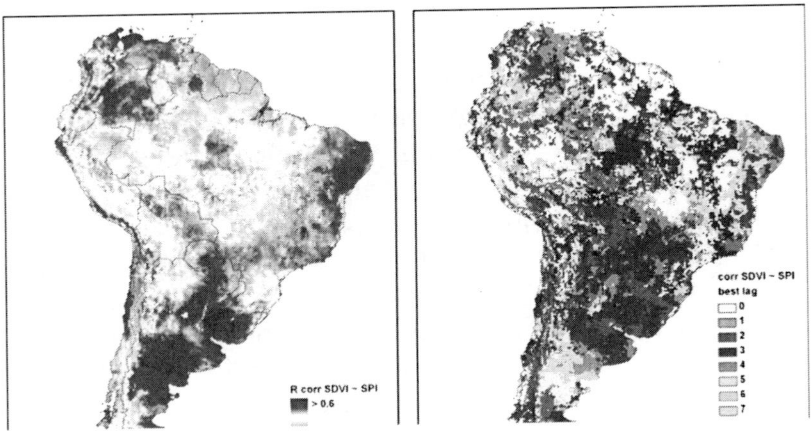

Figure 1: Correlation analysis between NDVI and SPI. Best lag expressed in months.

The results of the correlation analysis between NDVI and SPI are shown in Figure 1. Positively correlated (blue) areas suggest precipitation-vegetation coupling. In general, high positive correlation with best lags between 3 – 6 months are found in the semi-arid regions of South America, while weak (or even negative) correlation at all time scales was found in both the hot and humid zones and the deserts and high mountainous areas. These positive relationships between vegetation greenness and rainfall in drylands, where biomass production is determined by the amount of rainfall, and the opposite in humid and cold regions, where rainfall is not the limiting factor for vegetation growth, and deserts, where there is no rainfall at all, is consistent with findings of other authors, such as [33, 36].

In order to identify if a pixel is greening or degrading, and after identifying the best lag for each pixel, linear least squares trend analyses were performed on the SDVI time series and the SPI time series, taking into account the accumulated rainfall over the respective best lag. Only trends with Pearson correlation coefficients significantly different from zero (at significance level $p < 0.05$) are considered significant trends. Figure 2 shows the slope of the significant linear trends of vegetation greenness and precipitation anomalies, SDVI and SPI respectively. SDVI show slight but significant ($P<0.05$) positive trends in large areas in the northern part of South America, but significant negative trends in Argentina and the Peruvian coast. The results are comparable, but far more pronounced than results from annual series trend analysis, such as from [25, 28]. Also the SPI shows positive trends in the north-west of South America and the centre of Brazil. Negative precipitation trends are found in Argentina, the Peruvian coast and north-east of Brazil.

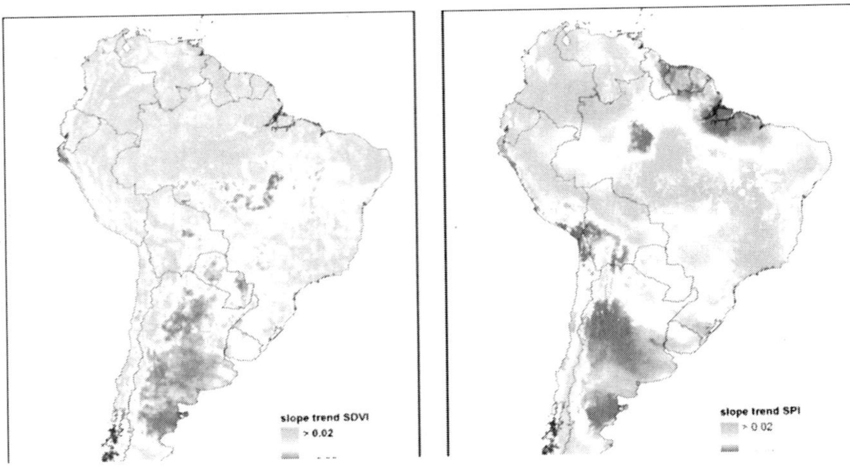

Figure 2: Slope of the trend analysis of SDVI (left) and SPI (right). Non-significant trends are masked in grey.

Following the trend analyses, a decision tree approach is adopted in order to interpret the results. Five queries are stipulated: (1) Does the time series of SDVI show a significant trend? (2) Are significant trends in SDVI coupled to significant correlations between NDVI and SPI? (3) Are significant trends in SDVI linked to significant trends in SPI? (4) Does the SDVI show a positive trend? (5) Do trends in SDVI correspond to trends in SPI?

The first question is meant to identify the focus area, i.e. pixels where the de-seasoned vegetation index (i.e. SDVI) shows a significant linear trend over time. The second and third queries are used to identify areas that differ in their relationship between vegetation and precipitation trends, respectively, while queries 4 and 5 are meant to distinguish positive and negative trends that are linked to trends in precipitation. On the other hand, it is possible to identify regions where positive or negative trends in vegetation are not linked to changes in precipitation and other climate variables or human impact play an imperative role. In step 4, pixels are divided in greening or degrading pixels. In the last step, both the trends in SDVI and SPI are evaluated.

The results of the decision tree analysis are shown in Figure 3. Green classes show a significant positive trend in SDVI coupled to an increase in SPI. Red classes show a coupling between a decrease in SDVI and SPI. The yellow and orange classes are classes where SDVI and SPI show an opposite trend. Argentina is clearly suffering from vegetation degradation linked to a decline in precipitation, which confirms the findings of [35]. The opposite is going on in Colombia and some parts of Brazil, where a process of greening seems to be linked to an increase in precipitation. Nevertheless, Brazil shows a patchy result, with some areas showing an increase in SDVI, although the precipitation decreases (yellow areas in Figure 3), and other areas showing a decline in vegetation, although precipitation increases (orange area in Figure 3), probably related to deforestation or forest degradation.

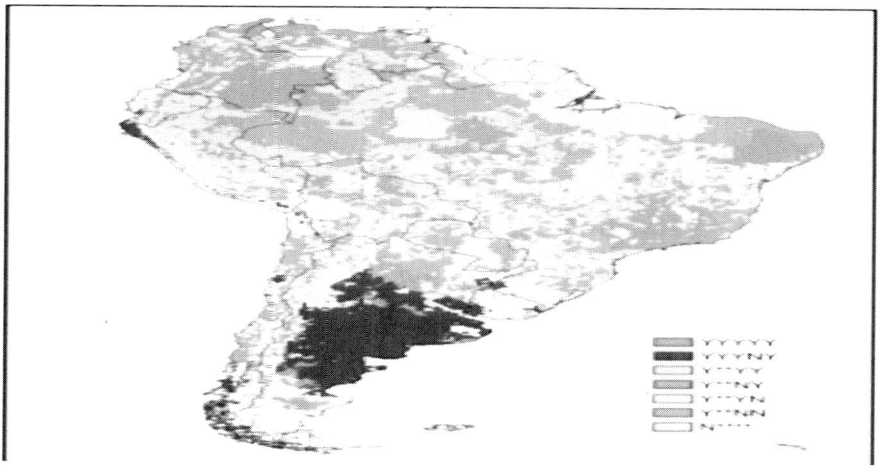

Figure 3: Result of the decision tree approach. Pixels are classified according to the 5 queries described in the text (where Y and N refer to a 'yes' or 'no' answer to the query, and * to either Y or N): green areas show an increase in SDVI, coupled to a significant positive trend in SPI (dark green) or not (light green); red areas show a decrease in SDVI, coupled to a significant negative trend in SPI (dark red) or not (light red); yellow areas show a positive trend in SDVI, but a negative trend in SPI; orange areas show a negative trend in SDVI, but a positive trend in SPI. Pixels without a significant trend in SDVI are masked in grey.

The resulting map can be used to estimate the coupling between vegetation (SDVI) and precipitation (SPI), shown in Figure 4. Three estimations are made: (A) a large estimation where significant trends in SDVI are coupled with trends in precipitation when SPI shows the same trend (significant or not); (B) an average estimation where significant trends in SDVI are linked to trends in precipitation when SPI shows a significant trend; and (C) a conservative estimation where significant trends in SDVI are linked to trends in precipitation when the SPI shows a significant trend and the correlation between NDVI and SPI is significant.

In estimation A, also weak trends in SPI are taken into account. These non-significant trends in precipitation are most probably not by itself responsible for significant trends in SDVI. Together with other variables like temperature change or human impact, these weak changes in precipitation might however give an extra impulse to greening or degradation processes. It is also possible that an increase in precipitation does not result in higher vegetation cover because the area is covered with climax vegetation or, the other way round, a decrease in precipitation does not result in further degradation because the area is already covered with minimal vegetation growth. Estimations A and B show little difference. The difference between estimation B and C is based on the significance of the correlation between NDVI and SPI, and shows larger differences, mainly in the greening pixels. In many pixels that show a positive trend in SDVI, vegetation greenness is not significantly correlated to SPI. It is therefore not certain that in these areas the increase in SDVI is coupled to an increase in precipitation.

From the conservative estimation, we can conclude that in 8% of South America, vegetation degradation is coupled to a significant decrease in the amount of precipitation in the last 14 years. Our results corroborate with the findings of [37]. In contrast, in 18% of the subcontinent, vegetation greenness has significantly increased over the last 14 years, coupled to an increase in precipitation. For 46% of the study area, significant degradation or greening processes could not be linked to changes in precipitation over time, indicating human impact or the influence of other climatic factors, such as temperature. Finally, and without taking into account the link with trends in precipitation, 40% of the subcontinent is showing an increase in photosynthetic activity over time, while desertification is taking place in 32% of the area.

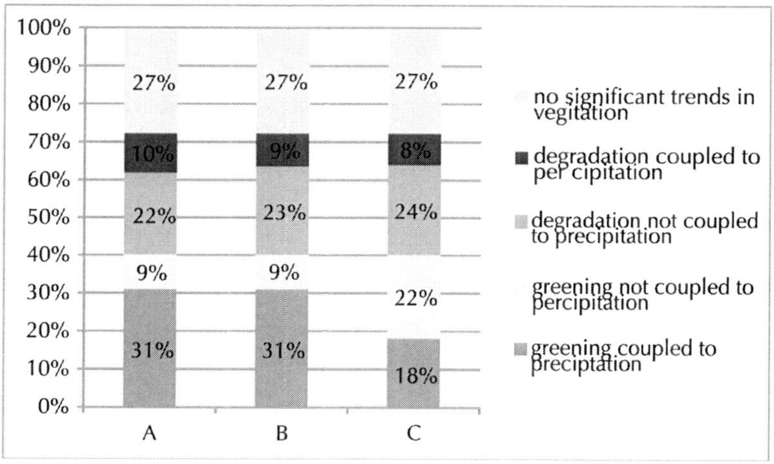

Figure 4: Large (A), average (B) and conservative (C) estimation of the linkage of greening and degradation to changes in precipitation over time. In case of estimation B and C, the yellow areas in Figure 3 are included in the light green fraction, while the orange areas in Figure 3 are included in the light red fraction.

TIME SERIES ANALYSIS FOR DROUGHT MONITORING

Drought is a recurrent feature of the Indian climate and usually begins at any season and can prolong for many years. The study of drought characteristics is to ascertain the spatial and temporal distribution of droughts, synoptic meteorological conditions associated. The definition of drought mainly depends on the precipitation deficiency. Studies on droughts in India have been reported by many scientists based on the rainfall anomalies over a particular region [38-40]. According to India Meteorological Department (IMD) guidelines, drought is defined as the consequent rainfall deficiency (below 19% of normal) for a period of 2 consecutive weeks. However, this criterion varies from country to country, based on the meteorological/climatological conditions such as percentage of moisture present, land topography etc.

Drought is of different types including meteorological droughts, agricultural droughts, hydrological drought and socioeconomic droughts. These are based on the variations of rainfall, crop water, surface water and economic conditions respectively. These droughts show variation with respect to the climatology which prevails of that region. The climatology of a region is a replica of the severity of drought. The climatology can be derived from the Thoronthwaite Climate System by taking the inputs of rainfall and water need. The model gives the amount of moisture (annual/seasonal) from which the classification of climate can be studied.

In view of the above, the present study focuses in obtaining the climates in different parts of Karnataka state which is located at the western half of the Deccan Palateau of India. The Moisture Index values which are the basis for delineating climatic type were compared with the AVHRR NDVI to understand the drought climatology in different test regions of Karnataka, India (Figure 5).

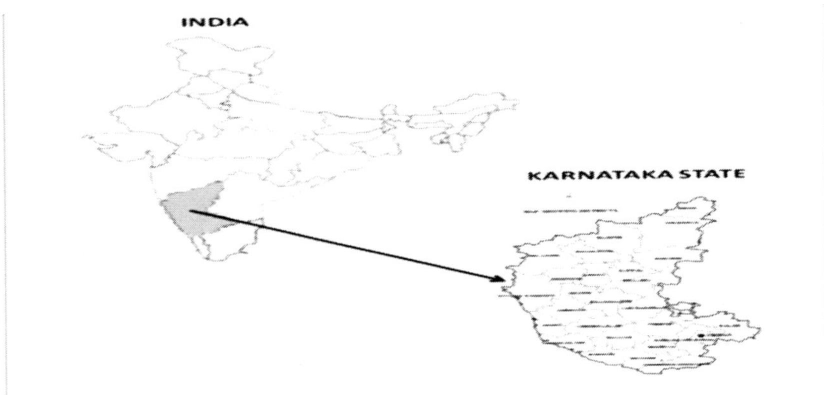

Figure 5: Study area of the present investigation.

The rainfall (P) and potential evapotranspiration (PE) data for the period 1982 to 2000 was downloaded from [41]. This data is based on the global rainfall and temperature (PE can be calculated from temperature) data sets of Climate Research Unit, University of East Anglia, United Kingdom. This was averaged for all the districts of Karnataka state till the year 2000 and uploaded website.

Taking the inputs of P and PE, we run awater balance model and derived the monthly Aridity Index (I_A) and Humidity Index (I_H). Moisture Index (I_M) which is the basis to tell the climatology of a region can be obtained by subtracting I_A from I_H. Table 1 below shows the infered climate types based on I_M as per Thoronthwaite Climate Approach [42, 43].

Table 1: Classification of climatic types based on Thoronthwaite Approach

Moisture Index(IM)	Climatic Type	Notation
100 & above	Perhumid	A
80 to 100	Humid	B4
60 to 80	Humid	B3
40 to 60	Humid	B2
20 to 40	Humid	B1
0 to 20	Moist Subhumid	C2
-20 to 0	Dry Subhumid	C1
-40 to -20	Semiarid	D
-60 to -40	Arid	E

The selected districts for the study are Chikkamagaluru, Belgaum, Chamrajnagar and Gulbarga of which climates are Humid, Dry subhumid, Semiarid and Arid. A comparative study was made with the seasonal values of IM and NDVI using time series and correlation analysis.

Table 2 shows the climatology of the four selected test sites during the period of 1982 to 2000. The overall climate of the test regions for the study period represented the humid, dry subhumid, semi arid and arid for Chikkamagaluru, Belgaum, Chamrajanagar and Gulbarga districts respectively.

Table 2: Climatic types of test regions from 1982 to 2000

Year	Climate Type			
	Chikkamagaluru	Belgaum	Chamarajanagar	Gulbarga
1982	C1	C1	D	E
1983	B1	C2	D	E
1984	C2	D	D	E
1985	C1	D	D	E

1986	C2	D	D	E
1987	C1	D	D	E
1988	B2	C1	D	E
1989	C2	D	D	E
1990	C1	D	D	E
1991	B1	C1	D	E
1992	B2	C2	D	E
1993	B1	B1	C1	E
1994	B2	C2	D	E
1995	C2	C1	D	E
1996	B1	D	D	E
1997	B1	C1	C1	E
1998	B2	C2	C2	E
1999	B1	B1	C1	E
2000	B1	C2	D	E

Chikkamagaluru (Humid Region)

The climatic types of Chikkamagaluru were dominated by humid type which is followed by the dry sub humid type. Chikmagalore (Humid region) recorded a maximum I_M of 450 during the year 1994 (Figure 6). The variation of I_M shows that it has increased from June to August in all years and recorded a comparative less value in the month of September. In a similar way, the NDVI progressed from June to September in all years with a maximum value of 0.55 for September, 1992. It is conspicuous that the variability in NDVI is more than I_M and the trends of I_M and NDVI were increasing. The correlation coefficient of these two indices is +0.08 which is very poor. The studies of [38, 39] also suggested that understanding the relation of NDVI with rainfall and its by-products is a very difficult task in humid regions. Since the plenty of moisture is already available in the soil, the vegetation can utilize the moisture for its growth and in such case it may not directly/ immediately dependant on rainfall derived indices in humid regions.

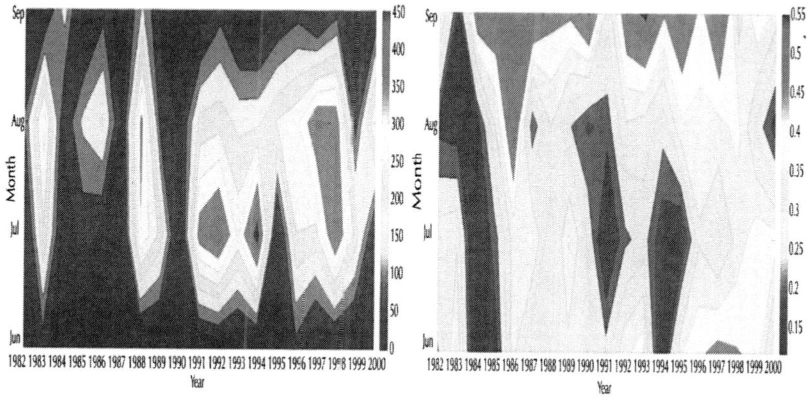

Figure 6: Variation of IM and NDVI – Chikkamagaluru.

Belgaum (Dry subhumid)

The climate of Belgaum varied from semi-arid to 1st humid type during the study period. 10 years of the study period show the subhumid climates followed by semi-arid in 7 years. The variation of I_M with NDVI over Belgaum unraveled the low variability of I_M associated with high variations of NDVI (Figure 7). The maximum NDVI is 0.55 that is during September of 1984 and 1992 linked with the very low values of I_M. The I_M values were around zero during all the years of study period in June and July months countered by the low values of NDVI. There is no sea – saw relation found between I_M and NDVI over this region. The correlation in this case was also found to be very poor and insignificant.

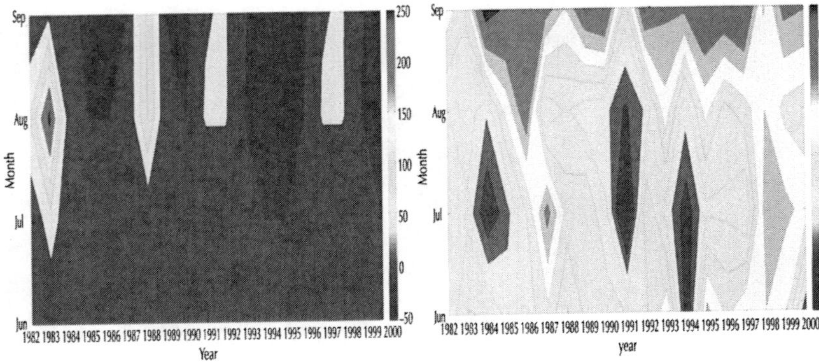

Figure 7: Variation of I_M and NDVI – Belgaum.

Chamrajnagar (Semiarid)

From Table 2, it is noticed that Chamrajanagar shows the arid climate category in 15 years of the study period. The years such as 1993, 1997, 1998 and 1999 displayed subhumid climates. The climatic types of this region are the good representative of climatic features over the study period. In the case of semi arid region of Chamrajnagar, the variability of I_M and NDVI is better as compared with the humid and dry subhumid regions (Figure 8). The overall I_M is varied from -10% in the year 1984 to a maximum of 44% during the year 1998 for the south west monsoon season. Accordingly NDVI, also varied from 0.26 to a high value of 0.42 in the year 1996. The values of NDVI are less in this region than previously mentioned areas but the trends of I_M are NDVI were positive with slopes of 0.824 and 0.002 for I_M and NDVI respectively along with the standard deviations of 11.5 and 0.05. The correlation of these two data sets is 0.42 at 0.05 level of significance which infers the good agreement of I_M and NDVI. From this analysis, it can be noticed that the deficiency of moisture which is represented by I_M in this study was well reflected by low NDVI values and adequate moisture conditions are supported by the moderate vegetation conditions.

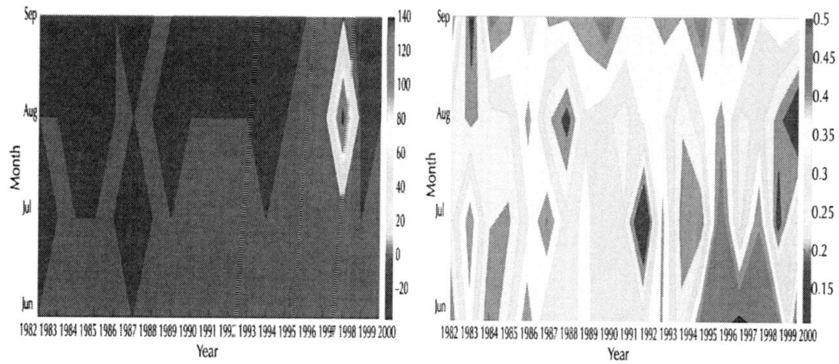

Figure 8: Variation of I_M and NDVI – Chamrajanagar.

Gulbarga (Arid)

All the years of study period are dominated by the arid category for this region. The comparison of I_M and NDVI yielded good results in the arid region of selected test sites (Figure 9). The interannual variability was found to be very high both in I_M and NDVI where as I_M varied from -77% during August of 1982 to zero value during September month in the years 1983 and 1992 with the corresponding NDVI values of 0.055 and 0.127 respectively. The time series plot for I_M and NDVI for the total south west monsoon season display the one to one linear agreement where the trends of both were highly increasing than other test regions (Figure 8). The slopes of the trends were 1.68 and 0.007 respectively. The standard deviations of 14 and 0.05 infer that the interannual variability of I_M is more than NDVI from which it can be noticed that the vegetation over a region may not respond immediately to the rainfall/available moisture despite there is a dependence of vegetation on rainfall/available moisture. The time series plot of I_M and NDVI (Figure 10) shows that the maximum of amount of moisture of-12% have seen in the year of 1998 with the NDVI value of maximum NDVI of 0.269 that is recorded during the entire study period. The correlation in this case is +0.64 which is at 0.01 level of significance which shows the strong agreement between I_M and NDVI.

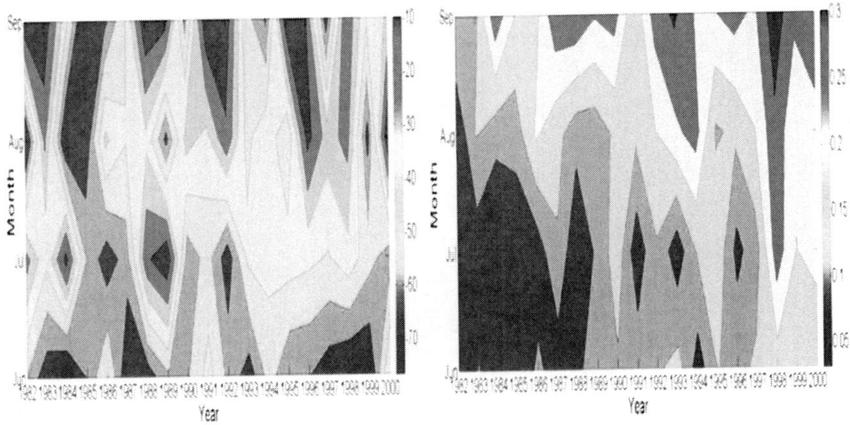

Figure 9: Variation of I_M and NDVI – Gulbarga.

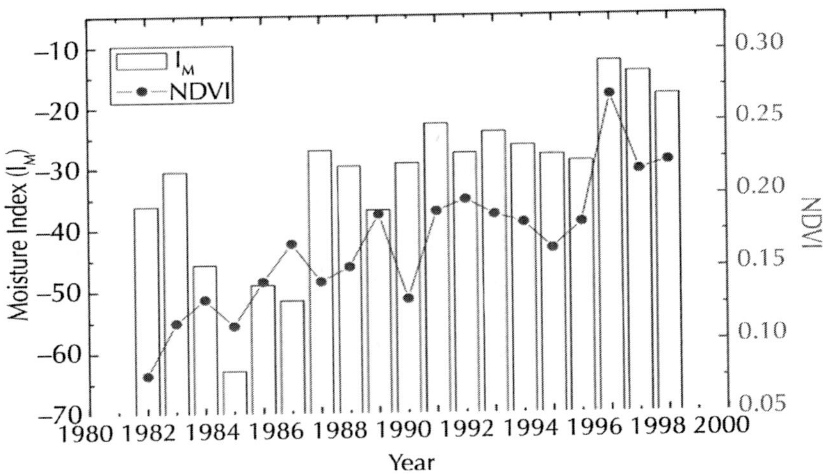

Figure 10: Time series of I_M and NDVI for southwest monsoon.

The study commenced with the retrieval of I_M values from the water balance model on monthly basis over different selected test regions those represent the various climatic types such as humid, dry sub humid, semi arid and arid. The knowledge of different climatic types enables us to understand the climatology of the test regions during the study period. Since these climates were derived from the $I_M$values, they replicate the status of the moisture content available which is very essential input to decide the crops fate. The comparison study of I_M with the satellite derived NDVI shown very interesting features of sensitivity of NDVI with I_M over different climatic types. The study inferred the poor correlation such that no linear and significant relation of I_M with NDVI over humid and dry subhumid regions. The reason for this could be the plenty of available moisture over these regions and even temporary perturbations of land surface conditions may not affect the crops/agriculture. The relation grown up to strong when the comparison closes from semi arid to arid regions. Especially, Gulbarga, arid region displayed very strong relation of I_M with NDVI which unraveled the poor/good vegetative conditions associated with low/high values of I_M. The correlation of +0.65 is a good supporting factor to say that the relation is substantial. The overall analysis of the present study suggested that the relation of I_M with NDVI is very strong and it is of immense use for the studies of drought monitoring in the arid areas as compared with the other climatic types.

QUANTIFICATION OF SUGARCANE CROP PRODUCTIVITY: A STUDY CASE IN SOUTHEASTERN BRAZIL

Agriculture represents an important segment of the economy of Brazil. Over the past 30 years, Brazilian agricultural growth and development has been guided by policies and technologies based on research for development. Remote sensed imagery plays an important role in agricultural crop production over large area, quantitatively and non-destructively, because agricultural crops are often difficult to access, and the cost of ground estimating productivity can be high. The recent development of GEONET Cast–EUMET Cast data has allowed us to obtain frequent and accurate measurements of a number of basic agrometeorological parameters (e.g. evapotranspiration, surface albedo, surface temperature, solar radiation, rainfall etc.). The GEONET Cast–EUMET Cast real-time and on-line data dissemination systems represent global network of satellite-based data dissemination systems designed to distribute space-based, air-borne and in situ data, metadata and products to diverse communities.

To determine agriculture productivity, this case study aimed to develop a GEONET Cast-EUMET Cast product-based method of estimating the productivity of sugarcane using an agrometeorological spectral model. The study was carried out in the Municipalities of Barretos and Morro Agudo, located in the state of São Paulo, Southeastern Brazil (Figure 11). The analysis was performed for 2009/2010 and 2010/2011 year's crop.

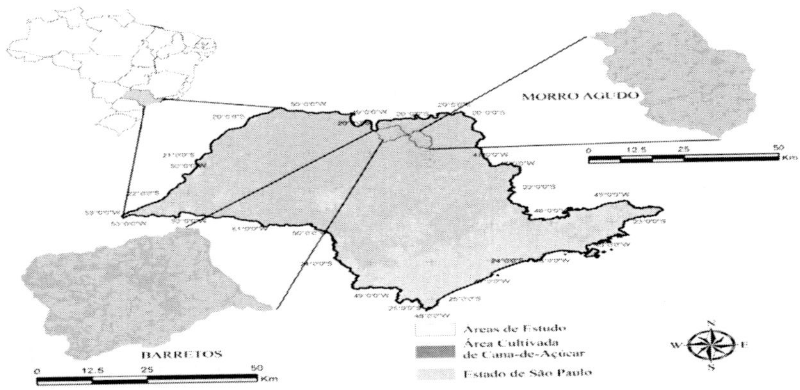

Figure 11: Spatial variability of crop yields (2010/2011) of Barretos and Morro Agudo in Sao Paulo, Brazil.

The values of sugarcane parameters used such as Respiration Factor (RF) (0.5 for temp. ≥ 20°C and 0.6 for temp <20°C); Agricultural Productivity Factor (APF) (2.9), Yield Response Factor (K_y) and Crop Co-efficient (K_c) were taken from [44-46]. The EUMETCast service is installed at Laboratory of Analysis and Processing of Satellite Images (LAPIS) at Federal University of Alagoas (UFAL). The remote sensing data of NDVI S10, Production of Dry Matter (DMP) are available at the website http://www.lapismet.com and SPOT Vegetation indices of VITO were collected form http://free.vgt.vito.be/.

This application proposes to test a remote sensing approach to quantify estimates of sugarcane productivity over the Coruripe municipality with the Integrated Land and Water Information System, (ILWIS, 3.7.1) GIS software. ILWIS was used to compute sugarcane crop estimates for each pixel in NDVI DMP and ET_p images by applying radiative, aerodynamic and energy balance physics in 7 computational steps. These images are currently provided over both daily and 10 day composites at about a 3km and 1km spatial resolution, by EUMETSAT and VITO respectively.

Step 1: Input NDVI and DMP Databases Using Algorithm Adapted from Geonetcast Toolbox

To implement a methodology for the ingestion of both the NDVI and DMP databases (raster) into ILWIS, specific routines of GEONETCast–toobox are adapted to import the datsets. For the ingestion procedure, based on a GIS approach using open source components, it requires additional work on corrections using overlays (Status map and LOG–file) to mask all appropriate areas of sugarcane crops over the Coruripe municipality.

Step 2: Computation of Fractional Vegetation Cover (FVC) from NDVI

For each pixel, the NDVI is converted to Fractional Vegetation Cover (FVC) by means of the the formula of [47]. The FVC is the one biophysical parameter that determines the contribution partitioning between bare soil and vegetation for surface evapotranspiration, photosynthesis, albedo, and other fluxes crucial to land–atmosphere interactions.

$$FVC = 1.1101 * NDVI - 0.0857.$$

Step 3: Computation of Leaf Area Index (LAI) from FVC

For each pixel, the FVC is converted to Leaf Area Index (LAI) by means of the formula of [48]. The LAI, defined, as the total one-sided leaf area per unit ground area, is one of the most important parameters characterizing a canopy. Because LAI most directly quantifies the plant canopy structure, it is highly related to a variety of canopy processes, such as evapotranspiration, interception, photosynthesis and respiration.

$$LAI = -2Ln(1 - FVC).$$

Step 4: Computation of Growth Factor from LAI

[49] developed a simple approach for deriving growth rate equation from LAI. Experimental evidence indicated that the growth rate of several agricultural crop species increases linearly with increasing amounts of LAI, when soil water nutrients are not limiting [46]. The following equation is used:

$$CGF = 0.515 - e[-0.667 - (0.515*LAI)]$$

where *CGF* = Corrected Growth Factor. Experimental evidence indicated that the growth rate of several agricultural crop species increases linearly with increasing amounts of LAI, when soil water nutrients are not limiting.

Step 5: Computation of Maximum Yield Potential (Y_p)

The final equation that was used to derive maximum yield potential (Y_p) includes evaporative fraction corrected growth factor (CGF), respiration factor (BF), agricultural productivity factor (APF) and production of dry matter (DMP) product.

$$Y_p = CGF*BF*APF*DMP$$

where Yp is the maximum yield potential (kg ha^{-1}).

Step 6: Retrieval of Evapotranspiration (Et_p) Via Land Surface Analysis –Satellite Application Facility (LSA SAF) Et_p Product

The crop coefficient is defined as the ratio of crop evapotranspiration, ET_r, to reference evapotranspiration, ET_p. K_c is crop specific and ranges from zero to over unity, depending on the crop growth stage. Crop evapotranspiration at any time during the growing season is the product of reference evapotranspiration and the crop coefficient.

$$ET_r = ET_p * K_c$$

Crop coefficients was developed for nearly all crops by measuring crop water use with lysimeters and dividing the crop water use by reference evapotranspiration for each day during the growing season of 2009/2010 [50].

The recent development of LSA–SAF products has allowed us to obtain frequent and accurate measurements of a number of basic agrometeorological parameters (e.g. surface albedo, surface temperature, evapotranspiration). The satellite estimated agrometeorological parameters have several advantages compared to conventional measurements of agrometeorological data in ground meteorological network.

Step 7: Estimation of Sugarcane Productivity

The sugarcane yield estimation model over the growing season, on a biweekly basis, is accomplished by using an agrometeorological model integrated to ILWIS according to [46]:

$$Y_e = Y_p[1 - ky(1 - ET_r/ET_p)]$$

where Y_e is the estimated yield (kg ha^{-1}), Y_p the maximum yield (kg ha^{-1}), k_y the yield response factor; ET_r the actual evapotranspiration (mm) and ET_p the maximum evapotranspiration (mm). Maximum yield (Y_p) is established by the genetic characteristics of the crop and by the degree of crop adaptation to the environment.

The resulting map of the estimated yield (*Ye*) is clipped to mask the Coruripe municipality boundaries in the State of Alagoas, Brazil. To establish correct coordinates, Map calculation within the ILWIS is used to implement this procedure. Flow diagram of methodology of quantifying sugarcane productivity via satellite products is shown in Figure 12.

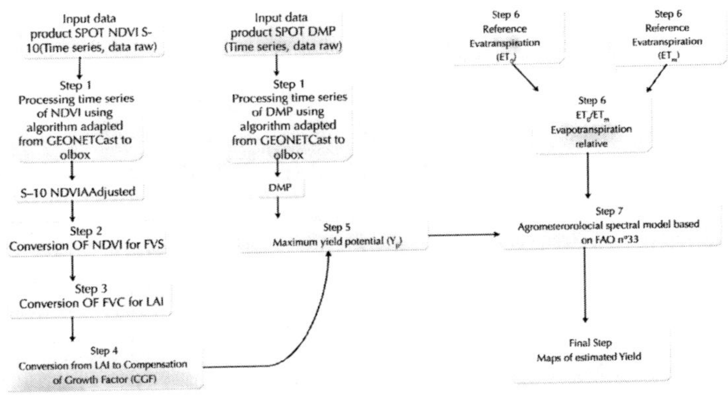

Figure 12: Flow diagram of methodology of quantifying sugarcane productivity via satellite products.

Figure 12 shows the spatial variations in sugarcane production over the Barretos and Morro Agudo municipalities for 2009/2010 and 2010/2011. The figure clearly indicates high spatial patterns in yield variability. This could be due to the mixing of significant fraction of observed pixels for the "arable pixel" and "non-arable pixel" within the municipalities. The quantified results give sugarcane yield mean range of 50 to 135 Ton ha^{-1}.The results obtained here represents a first step towards an operational use of ILWIS tools in Brazil using NDVI S-10, DMP SPOT and ET_o for operational estimating of sugarcane productivity. Overall, the model was able to identify (Figure 12) and quantify (Table 3) the spatial variability of agricultural production over the municipalities analysed. Therefore, the methodology is useful for developing estimates of operational support for the sugarcane productivity [51].

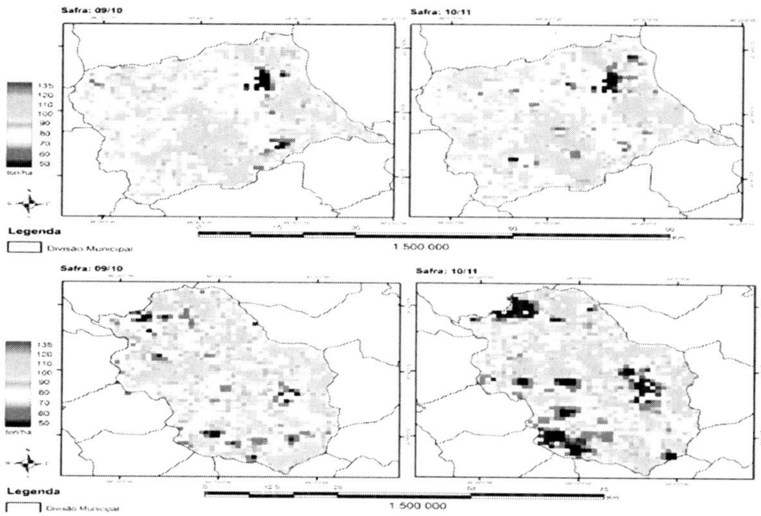

Figure 13: Spatial variability of crop yields over the Barretos and Morro Agudo municipalities for 2009/2010 and 2010/2011.

Table 3: Comparison between the productivity of sugarcane using an agrometeorological spectral model and harvested crop yield from National Food Supply Company (CONAB)

Mean Productivity (t/ha)		
Barretos	Crop	
	2009/2010	2010/2011
Agrometeorological model	93,15	93,31
CONAB	96,21	93,29
Mean Productivity (t/ha)		
Morro Agudo	Crop	
	2009/2010	2010/2011
Agrometeorological model	87,96	84,10
CONAB	93,60	87,25

TIME SERIES ANALYSIS FOR AGRICULTURE MONITORING: UGANDA

The economy of Uganda and its development goals are heavily premised on agriculture. Over 79% of the households are engaged in agriculture while 73% are directly or indirectly employed in the agricultural sector. Uganda's agriculture is however almost entirely rain-fed and very susceptible to climate risks. Studies indicate that Uganda's agricultural sector will be adversely affected by climate variability and projected climatic changes making real time monitoring of crop growth and crop productivity very important for better adaptation to climate variability and climate change. Quantitative analyzes reveal that the agricultural sector in Uganda needs to grow at an annual rate of 7% to effectively contribute to national development. Currently, the rate of agricultural growth in Uganda is below the population growth. With the threats of climate change and variability, Uganda needs to among other things harness geo-information technologies (remote sensing and geographical information systems) to improve agricultural productivity.

Remotely sensed images are powerful tools in monitoring crop productivity and yields. In most developed countries and emerging developing countries like Brazil and India, remote sensing has been greatly harnessed to plan for agriculture production, monitor crop growth and estimate yields. This is paramount in the sense that timely interventions can be taken and obviates possibilities of famine and food insecurity. Although there have been strides taken to improve the utility of remote sensing in the agriculture some developing countries, a lot remains to be done to make it more efficient, relevant and more productive. An investigation of the causative factors of the low utility and uptake of remote sensing in the agricultural sector in Uganda implicates a number of factors ranging from low capacities to expensive images. Recent developments have however extended numerous opportunities in utilizing remote sensing in the agricultural sector.

The onset of utilization of remotely sensed techniques in Uganda was in the early 1990s spearheaded by the National Biomass Project and focused largely on land use and land cover mapping. The activities of the National Biomass Project were later taken on by the National Forestry Authority (NFA) but the domains and scope remained largely the same with more focus on land use, land cover and related aspects being given priority. Apart from the NFA, academic institutions of higher learning and to some extent some research institutions like the National Agriculture Research Laboratory Kawanda (NARL) and the National Environment Management Authority (NEMA) have some remote sensing application either for teaching or research. In general, the remote sensing applications in Uganda in the agricultural field are scattered and more project based. This is partly due to the fact that there is lack of a government agency with a clear mandate to spearhead and propel the utility of remote sensing application in the country. Nevertheless, some efforts through the government cooperation with UN agencies such as the FAO regularly provide some information analyzed at the regional level for early warning in the agricultural sector. Some of the historical constraints to efficiently harnessing remote sensing in natural resource management in Uganda are generally those also experienced in other developing countries in Africa including the high costs of imagery data, processing software, coarse resolution of images. inadequate physical and human capacities and weak institutions. To-date, most of the issues to do with data costs and software have been significantly resolved with many freely available images, open source versatile software or special low for developing countries on commercial software. The contemporary challenge now is more of institutional/agency capacities, human capacities and policy environment for enhancing the utilization of remote sensing in the country.

A range of great opportunities, hitherto unavailable exist now for effectively using remote sensing in agriculture and natural resource management, notably through; (a) datasets disseminated through the Geonetcast Platform (b) freely available and downloadable datasets (c) open source softwares and low cost commercial softwares. Details of the Geonetcast is fully described in various sources (e.g. [48-50]). In brief, its a low cost facility which enables dissemination of near real time satellite imagery data. It is part of the emerging Global Earth Observation system of Systems (GEOSS), led by the Group on Earth Observation (GEO), for environmental analysis [54]. The Geonetcast does not require internet connectivity which is always a major constrain in developing countries and the data is disseminated at a very high temporal resolution through a ground recieving station, making monitoring easy. The facility streams diverse datasets which can broadly be used in environmental monitoring covering agriculture, water, soils, fire forestry etc. The data can be processed using the ILWIS software, where a specific toolbox has been developed.

In this case study, we demonstrate the utility of relatively low spatial but high temporal resolution satellite images from earth observation systems in monitoring and assessment of agricultural productivity in Uganda. There are two main inputs i.e. production data and remotely sensed data. Production data was obtained online from the FAOSTAT [55]. We extracted the annual yield, production and harvested area of the top five crops produced in Uganda according to FAOSTAT; (1) plantain/banana (2) cassava (3) sweet potatoes (4) sugarcane and (5) maize for 10 years spanning from 2001 to 2010. Bananas/plantain (*Musa spp*) are largely grown in central, western and eastern (highland areas) parts of Uganda. As perennial crops, banana are year round crops. Cassava (*Manihot esculenta*) is an important food security crop in Uganda with the largest production coming from eastern and northern Uganda. Cassava accounts for approximately 13% of the daily caloric intake in Uganda. Cassava is commonly planted in the first season which is around February-march in most parts of the country and its also a perennial crop. Sweet potatoes (*Ipomoea batatas*) are also a food security crop in Uganda grown largely in the mid to high altitude regions of the country (1000-3000 meters above sea level). They are annual crops grown twice a year with the first season between February and June, while the second season stretches from September to November. They thrive well in the deep volcanic soils of Southwesten Uganda and Eastern Uganda. Sugarcane (*Saccharum officinarum*) in Uganda is largely grown on large plantations mainly in the near east and western Uganda. There are also a couple of out growers who are supported by sugar companies. It is mainly an income generating perennial crop. Maize (*Zea mays L.*) is grown in almost every part of the country and is a major staple food crop. It is an annual crop grown twice a year (March to June and September to November) in areas of the country where biophysical conditions are supportive. To ease the analysis, the production and yields for the five crops were compounded into one annual value.

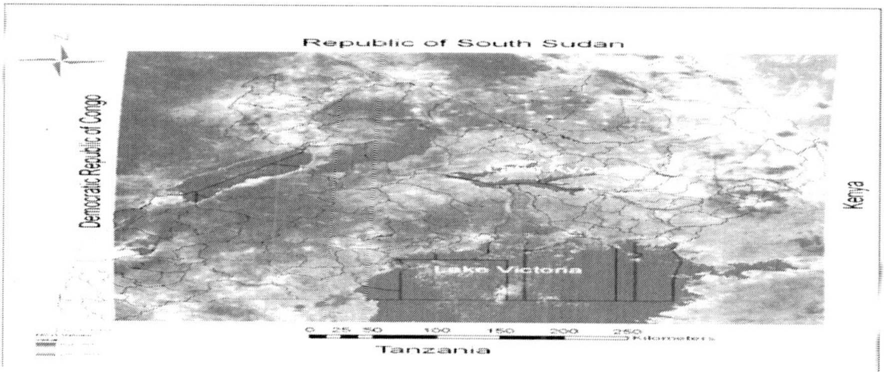

Figure 14: Scope of study.

The results on harvested area, production and yields are shown in Figures 15, 16 and 17 respectively.

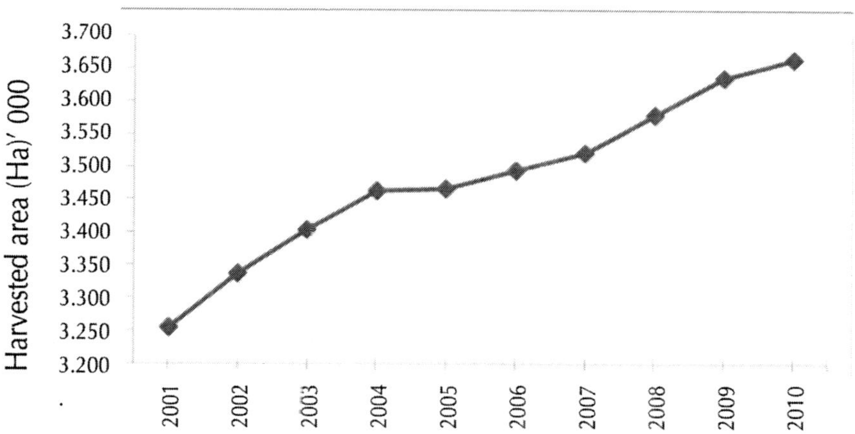

FIGURE 15: Harvested area of five crops between 2001 and 2010.

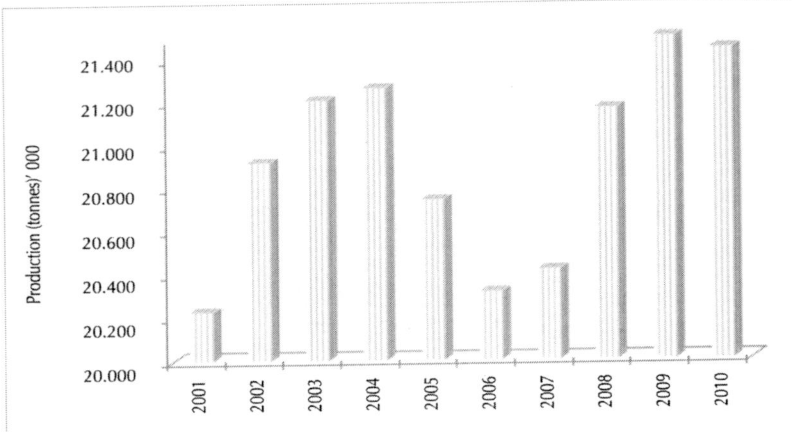

Figure 16: Production (compunded) trends of five selected crops between 2001 and 2010 compounded.

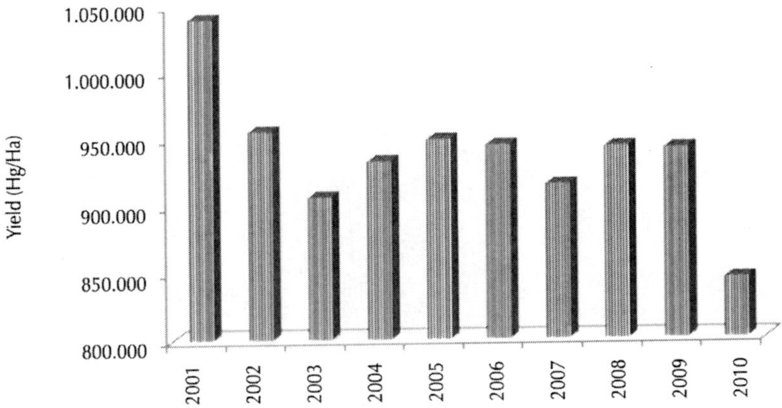

Figure 17: Yields (compounded) of five major crops between 2001 and 2010.

The results based on the three factors i.e. area harvested, production and yields do not depict a definitive trend. In terms of harvested area, there is an increasing trend, implying that more areas are being converted for cultivation of the specified crops (Figure 15). Production between 2001 and 2010 has modestly increased. Yields per hectare are however more variable and actually show and generally declining trend. Bearing in mind that production is increasing, it becomes explicit that the increments in production are related to extensification rather than intensification. In most cases, extensification entails conversion of ecologically sensitive and fragile areas such as wetlands or reclamation of forest area which has its environmental implications. Subjected to a statistical analysis, the results revealed a strong and positive correlation between the yields and production area ($r^2=0.52$, $p<0.05$).

Remote sensing analysis was on the MODIS NDVI data, which has a spatial and temporal resolution of 250 m and 16 days respectively. For each year 23 images are available in a decal arrangement. For the 10 year period, we downloaded a total of 230 images in HDI format and processed them in ERDAS Imagine where file format conversions were undertaken and later ILWIS for arithmetic analysis. Individual images (23 decades) for each year were stacked to generate a single profile for each year. Relevant statistics such as the mean, standard deviation, coefficient of variation were late extracted. The spatial distributions of average NDVI for selected years are shown in Figure 18, whileFigure 19 gives the temporal average NDVI dynamics for the 10 years. Mean average NDVI value is 0.56. In spatial terms, the southern part of the country registers higher NDVI values than the northern part. This is not surprising in light of the coverage in terms of natural cover and the crops grown which significantly entail banana and a range of annual crops grown in two seasons. The North Eastern part is particularly poorest in terms of annual average NDVI values. Understandably it is a semi arid region and generally more tailored to livestock enterprises than cropping enterprises. Annual NDVI values for the whole country were subjected to a correlation analysis with production and yield data, resulting into poor and insignificant correlations ($r^2=0.19$ to 0.2.1). The low coefficient are partly explained by the fact that some crops like sugarcane are eother irrigated or are grown in areas almost permanently under water (wetlands). However when the data was collapsed into the growing seasons and the water bodies excluded from the analysis, better and significant correlations were obtained (r^2 0.46 to 0.61, $P<0.05$) demonstrating the efficacy of using NDVI for crop monitoring and yield prediction. In light of the expected variability and changes in climate, coupled with the availability of data in real time, the NDVI analysis represents a great potential in sustainable adaptation where from both a policy perspective and direct intervention. This has positive implications in timely provisioning of information to farmers, adaptation to climate change and variability as well as enabling science based policy options for appropriate interventions. Specific prediction coefficients for different crops and regionalized to the climatic conditions can be helpful to local governments where timely interventions can obviate social instability related to crop failures. On the other hand, predictions of higher yields can also enable relevant agencies to solicit for markets for the produce, improving the welfare and livelihood of the farmers, who in the context of Uganda are largely small holder farmers. All these can only be realized if there is a good policy framework that ties all the relevant pieces in the chain i.e. science, production, markets and institutions.

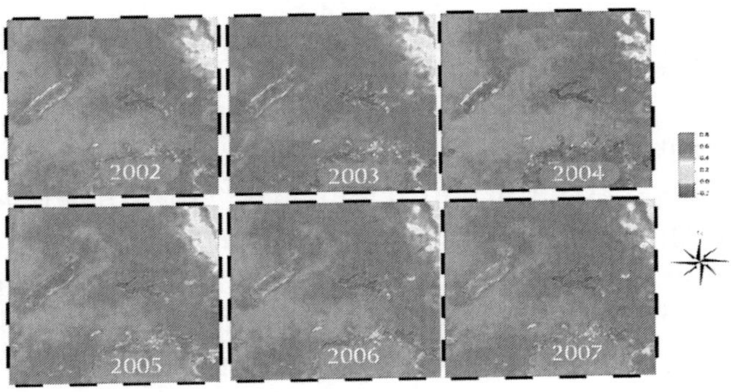

Figure 18: Average annual NDVI for selected years.

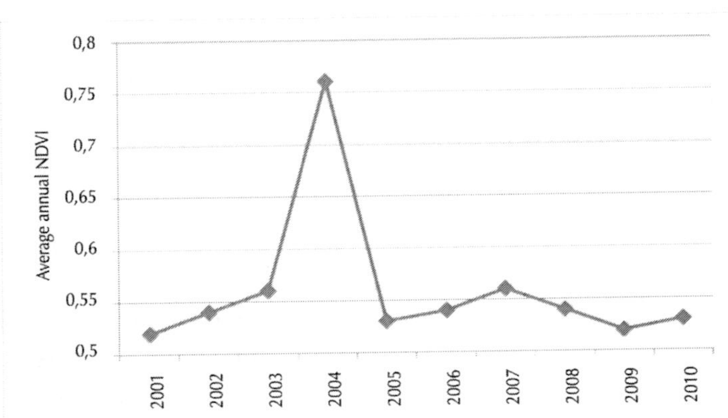

Figure 19: Variation of annual NDVI values between 2001 and 2010.

OUTLOOK AND CONCLUSIONS

We have provided four multi-disciplinary case studies on the power of using remote sensing technologies, and more specifically time series analysis of low resolution satellite derived indicators, for monitoring and analysing land cover changes, desertification, drought and agricultural activities on different spatio-temporal scales. Generating this information at finer temporal resolutions is crucial for reducing risks to disaster, preparedness and formulation of strategies for better adaptation to climate change particularly the increasing dramatic hydro-meteorological events in developing and emerging countries.

This chapter provides a variety of methodologies of processing chains over satellite data, allowing the monitoring of areas subject to or in risk of desertification and land degradation processes. This chapter provides new insights related on the use of remote sensing data for climate (change) impact monitoring, which will contribute to the advance of warning systems and adaptation measures in developing and emerging countries. The focus of future activities should however focus on institutional support and capacity building for impact assessments for Africa, South America and India. The importance of training and joint cooperation with local providers and users cannot be overestimated.

One of the most robust, multi-purpose and yet simple remote sensing index is the NDVI. NDVI imagery data is widely available for immediate use at almost no cost. This has been given emphasis in this chapter through demonstration of its utility in various environmental and production domains. The section 2 of the chapter mainly emphasizes the relation between trends of vegetation greenness and rainfall over a long term period, taking into account the time lag between rainfall and vegetation response. As a result, areas of greening or degradation can be identified, and the process can be linked or not to changes in precipitation.

The section 3 of the chapter tells the mode of relation between NDVI and moisture index over different climatic regions. The relation was found to be poor over humid and dry subhumid regions where as it is improving in semi arid and arid regions. The relation of above cannot be taken as granted in the humid regions though it is implicitly understood that NDVI maintains positive relation with IM. The study infers that the NDVI and IM relations cannot be used to characterize the drought over humid regions but can be taken as an indicator in arid and semi arid regions. This is particularly relevant for adaptation purposes in semi arid regions which cover big chunks in Africa, India and some parts of Southern America.

Section 4 of the chapter mainly focuses on the estimation of sugar cane yields in Southeastern Brazil by using spatial tools which have been integrated in ILWIS 3.7.1, open source software. This study underpins that the NDVI data along with the other meteorological data is of immense use for the estimation of crop yields. This gives a business orientation on the utility of spatial tools, but also has a livelihood implication where small scale farmers or out growers are involved in sugarcane production. Interestingly, sugarcane is a major crop in all the case study countries in this chapter.

The last section of the chapter also gives more emphasis on yield estimates of five major crops in Uganda. The results of the study showed that the production between 2001 and 2010 has modestly increased with the variability in yields. Also, this analysis showed that the extensification of crops is dominated by intensification and it is implied that the increments in production are related to extensification.

In a nutshell, the chapter demonstrates how remotely sensed data available in the public domain freely or at very low cost can be harnessed to address critical challenges in developing countries pertaining to environment, agricultural productivity, drought, desertification and ultimately climate change adaptation. The chapter shows that relating the satellite derived vegetation indices with existing models and parameters can be useful proxies to understand the various phenomena of the crops. However, despite the availability of the technology, full benefits from available remotely sensed imagery resources for developing countries can only be realized when enabling policies are formulated and implemented and concerted capacity development is undertaken to establish a critical human resource base. This will enable the policy makers to go for the risk managing practices such as agricultural crop reinsurance schemes, drought defining criteria etc.

In light of the resource constraints in developing countries, cooperation and collaboration is important to develop a nucleus of future demand and contributing to new scientific insights related to projected changes in drought drawing information from satellite data, which will contribute to the improvement of warning systems and adaptation measures in developing and emerging countries.

REFERENCES

1. L Wolf, and M Williams, GEONETCast- Delivering environmental data to users worldwide (September 2007IEEE Systems Journal 2008;23401405
2. UNEPClimate change strategy, United Nations Environment Programme 2008
3. IPCCClimate Change 2007Synthesis Report- Contribution of Working Groups I, II and III to the Fourth Assessment Report of the Intergovernmental Panel on Climate Change, Geneva, Switzerland 2007.
4. G Nelson, M Rosegrart, J Koo, R Robertson, T Sulser, T Zhu, C Ringler, S Msangi, A Palazzo, M Batka, M Magalhaes, R Valmonte-santos, M Ewing, and D Lee, Climate Change: Impact on Agriculture and Cost of Adaptation," IFPRI, 200919
5. Barbosa H.A; Lakshmi Kumar2011Strengthening regional capacities for providing remote sensing decision support in drylands in the context of climate variability & change. In: Young, S. S.; Silvern, S.E. Environmental change. USA, 979-9-53307-109-0
6. C Vörösmarty, P Green, J Salisbury, and R Lammers, Global water resources: vulnerability from climate change and population growth. Science 2000289284
7. WMO2011Climate Knowledge for Action: A Global Framework for Climate Service- Empowering the Most Vulnerable. World Meteorological Organization, WMO-1065
8. P Sellers, Canopy reflectance, photosynthesis and transpiration. International Journal of Remote Sensing 19856813351372
9. Sarma A.A.L.N & Lakshmi Kumar, T.2006*Studies on crop growing period and NDVI in relation to water balance components*, Indian Journal of Radio & Space Physics, 35, 424-434.
10. C Bhuiyan, F. N Kogan, 2008*Monsoon dynamics and vegetative drought patterns in the Luni basin under rain-shadow zone*, Internation Journal of Remote Sensing, 3132233242

11. Lakshmi Kumar T.V., Humberto A. Barbosa, K. Koteswara Rao & Emily Prabha Jothi(2012*Studies on the frequency of extreme weather events over India*, Journal of Agriculture Science & Technology, 1413431356

12. J Cihlar, St. Laurent, L., & Dyer, J.A., (1991*The relation between normalized difference vegetation index and ecological variables*, Remote sensing of Environment, 35279298

13. M. L Davenport, S. E Nicholson, 1993*On the relation between rainfall and the Normalized Difference Vegetation Index for diverse vegetation types in East Africa*, International Journal of Remote Sensing, 14, 2369.

14. H. A Barbosa, M. d. S. & T. V Mesquita, Lakshmi Kumar (2011*What do vegetation indices tell us about the dynamics of the Amazon evergreen forests?*, Geophysical Research Abstracts, 13, EGU 2011, 12894.

15. F. N Kogan, 1997*Global Drought Watch From Space*. Bulletin of the American Meteorological Society, 78, 621636

16. L. S Unganai, and F. N Kogan, 1998*Southern Africa's Recent Droughts From Space*.Advance in Space Research, 21, 3507511

17. P. S Ramesh, R Sudipa, and F Kogan, 2003*Vegetation and temperature condition indices from NOAA AVHRR data for drought monitoring over India*. International Journal of Remote Sensing, 24, 43934402

18. Z Wan, P Wang, and X Li, 2004*Using MODIS land surface temperature & normalized vegetation index for monitoring drought in the sourthern Great Plains, USA*, Int. J. Remote Sens., 256172

19. J. K Knight, R. L Lunneta, J Ediriwickerma, S Khorram, 2006*Regional scale land cover characterization using MODIS-NDVI. 250 m multi- temporal imagery: A phenology based approach*. GIScience and Remote Sensing (Special issue on Multi- Temporal Imagery Analysis), 43(1), 1-23

20. Chris Funk & Michale E. Budde (2009*Phenologically- tuned MODIS NDVI based production anomaly estimates for Zimbabwe*. Remote Sensing Environment. 113115125

21. F Lupo, I Reginster, and E Lambin, Monitoring land-cover changes in West Africa with SPOT Vegetation: impact of natural

disasters in 1998-1999. International Journal of Remote Sensing 2001221326332639

22. W Immerzeel, R Quiroz, and S De Jong, Understanding precipitation patterns and land use interaction in Tibet using harmonic analysis of SPOT VGT-S10 NDVI time series. International Journal of Remote Sensing 2005261122812296
23. J Xiao, and A Moody, Geographical distribution of global greening trends and their climatic correlates: 1982-1998. International Journal of Remote Sensing 2005261123712390
24. A Anyamba, and C Tucker, Analysis of Sahelian vegetation dynamics using NOAA-AVHRR NDVI data from 1981-2003. Journal of Arid Environments 2005633596614
25. U Helldén, and C Tottrup, Regional desertification: a global synthesis. Global and Planetary Change 200864169
26. Z Bai, D Dent, L Olsson, and M Schaepman, Global Assessment of Land Degradation and Improvement- 1. Identification by Remote Sensing, Report 2008/01, ISRIC- World Soil Information, Wageningen 2008
27. E Lepers, E Lambin, A Janetos, R De Fries, F Achard, N Ramankutty, and R Scholes, A synthesis of information on rapid land-cover change for the period 1981-2000. BioScience 2005552115124
28. G Baldi, M Nosetto, R Aragón, F Aversa, J Paruelo, and E Jobbágy, Long-term satellite NDVI data sets: evaluating their ability to detect ecosystem functional changes in South America. Sensors 200885397
29. F Achard, H Eva, H-J Stibig, P Mayaux, J Gallego, T Richards, and J-P Malingreau, Determination of deforestation rates of the world's humid tropical forests. Science 2002297999
30. UNCCDDesertification: a visual synthesis, United Nations Convention to Combat Desertification (UNCCD) Secretariat, Bonn 2011
31. C Tote, K Beringhs, E Swinnen, and G Govers, Monitoring environmental change in the Andes based on SPOT-VGT and NOAA-AVHRR time series analysis. In 6th International workshop on the Analysis of Multi-Temporal Remote Sensing Images, 2011
32. A Klisch, A Royer, C Lazar, B Baruth, and G Genovese, Extraction of phenological parameters from temporally smoothed vegetation indices. 2006

33. Z Bai, and D Dent, Global assessment of land degradation and improvement: pilot study in Kenya, Report 2006/01, ISRIC- World Soil Information, Wageningen 2006
34. A Peters, E Walter-shea, L Ji, A Vina, M Hayes, and M Svoboda, Drought monitoring with NDVI-based standardized vegetation index. Photogrammetric Engineering and Remote Sensing 20026817175
35. N Guttman, Accepting the standardized precipitation index: a calculation algorithm. Journal of the American Water Resources Association 1999352311322
36. A Lotsch, M Friedl, B Anderson, and C Tucker, Coupled vegetation-precipitation variability observed from satellite and climate records. Geophysical research Letters 2003
37. Z Bai, and D Dent, Land Degradation and Improvement in Argentina- 1. Identification by remote sensing, ISRIC- World Soil Information, Wageningen 2007Thoronthwaite and Mather, 1955)
38. Lakshmi Kumar T.V., Uma, R, Aruna, K & Emily Prabha Jothi(2011On the relation of NDVI with rainfall in Western Ghats, India, RAISE Conference Proceedings, 4245Dec 2011.
39. Lakshmi Kumar T.V., Uma, R, Koteswara Rao K & Humberto Barbosa(2012Variability of NDVI in relation to southwest monsoon over Western Ghats, India, International SWAT Conference, July 2012.
40. Liu Xiaomang Dan Zhang., YuzhouLuo&Changming Liu(2012Spatial and temporal changes in aridity index in northwest China: 1960 to 2010, Theoretical and Applied Climatology, DOIs00704-012-0734-7, Aug 2012.
41. www.indiawaterportal.org
42. Thoronthwate C W & Mather J.R(1955The water balance model, Publ in Clim, Drexel Institute, 8,1, 104
43. C W Thoronthwaite, 1948An approach toward a rational classification of climate, Geographical Review, 38(1), 55-94.
44. J. R. F Gouvea, 2008Mudancas Climaticas e a expextiva de seus impactos da cultura de cana- de acucar na regiao Piracicaba- *SP*, Dissertacao (Mestrado em Fisica do Ambiente Agricola)- Escola Superior de Agricultura " Luiz de Querioz", Universidade de Sao Paulo, Piracciaba, 99.

45. B. F. T Rudorff, 1985Dados Landsat na estimativa da produtividade agrícola da cana-de- açúcar.114, Dissertação (Mestrado em Sensoriamento Remoto)- Instituto Nacional de Pesquisas Espaciais- INPE, São José dos Campos.

46. J Doorenbos, A. H Kassam, 1979Yield response to water. Roma: Food and Agriculture Organization of the United Nations, 193.

47. J. C. J Munoz, J. A Sobrino, L Guanter, J Moreno, A Plaza, P Matinez, 2005Fractional Vegetation Cover Estimation from Proba/ CHRIS Data:Methods, Analysis of Angular Effects and Application to the Land Surface Emissivity Retrieval. 3rd ESA CHRIS/PROBA Workshop, ESRIN, ESA SP-593, Frascati, Italy.

48. J. M Norman, M. C Anderson, W. P Kustas, A. N French, J Meicikalski, R Torn, G. R Daik, T. J Achmugge, Remote Sensing of Evapotranspiration for Precision-Farming Applications. In: IEEE International Geoscience and Remote Sensing Symposium, 2003Tolouse. Proceedings: IGARSS 2003OrganizingCommittee, 21-25.

49. L. M. S Berka, B. F. T Rudorff, Y. E Shimabukuro, Soybean yield estimation by an agrometeorological model in a GIS. Scientia Agricola, 60n. 3; 433-402003

50. Toledo Filho M. R.1988Probabilidade de suprimento da demanda hídrica ideal da cultura de cana- de-açúcar (Saccharum spp.) através da precipitação pluvial na zona canavieira do estado de Alagoas. 1988.72f. Dissertação (Mestrado)- Progama de Pós-graduação em Agronomia, Escola Superior de Agricultura "Luiz de Queiroz", Universidade de São Paulo, Piracicaba.

51. Barbosa H. A; Rocha, D; Lakshmi Kumar, T.V., Bamutaze, Y.Quantification of sugarcane crop productivity using agrometeorological-spectral model: A study case in Northeastern Brazil. Revista AGROLLANIA, 1690-80669Enero-Diciembre, 62-75, 2012

52. http://www.earthobservations.org/geonetcast.shtml,

53. http://www.eumetsat.int/Home/Main/DataAccess/EUMETCast/ SP_20100519114624675?l=en,

54. http://www.itc.nl/Pub/WRS/WRS-GEONETCast/GEONETCast-toolbox.html http://faostat.fao.org/

Chapter 2

Multi-Scale GIS Data-driven Method for Early Assessment of Wetlands Impacted by Transportation Corridors

Rodrigo Nobrega[1], Charles O'Hara[1], Bethany Stich[1,] and Colin Brooks[2]

[1]Geosystems Research Institute, Mississippi State University, USA
[2]Environmental Science Laboratory, Michigan Tech Research Institute, USA

INTRODUCTION

The correlation between transportation systems and adverse impacts on the natural environment have been investigated at different scales of observation (Kuitunen et al., 1998; Bouman et al., 1999;Corrales et al.,

2000; Formann et al., 2003, Wheeler et al., 2005, Fletcher and Hutto, 2008). There is a growing body of literature reporting and quantifying the effects caused by transportation infrastructure on the proximate biophysical setting as shown in (Keller & Largiardèr, 2003) as well as on the socio-economic setting as shown in (Boarnet & Chalermpong, 2001). The environmental consequences of landscape fragmentation in different phases of transportation project development have been investigated and tabulated by (Corrales et al., 2000). However, the disparity of definitions for the biophysical landscape can make it difficult to communicate clearly and even more difficulty to establish consistent management policies. Landscape invariably comprises an area of land containing a mosaic of patches or land elements (McGarigal & Marks, 1995; Hilty et al., 2006). The overall knowledge-base of transportation systems and methods to consider, minimize, and mitigate adverse impacts on natural systems and biophysical settings have gradually been absorbed and adopted by transportation and Environmental Impact Assessment (EIA) practitioners to design balanced engineering solutions and deliver transportation infrastructure in an environmentally responsible manner. The body of science and knowledge supporting practitioners has grown through in-depth reviews about transportation and ecological effects (Spellerberg, 1998; and Formann et al., 2003) Similarly, the knowledge base concerning the impacts of land use on travel behaviour is also being investigated and developed from the transportation perspective (Mokhtarian & Cao, 2008; Litman, 2008).

Road development is a primary mechanism responsible for habitat, ecosystem, and overall biophysical fragmentation, replacing or modifying pre-existing land cover such as wetlands, creating edge habitat and altering landscape structure and function (Saunders et al., 2002). While conserving the remaining natural environment as well as restoring environmentally impacted areas is vital for natural sustainability, transportation corridor development is required by society and results in our modern transportation infrastructure and travel patterns.

Previous lessons learned show that environmental issues should be considered early the transportation planning process in order to balance economic, engineering and natural sustainability perspectives (Amekudzi & Meyer, 2006). A highway design that meets the transportation corridor needs, while minimizing environmental impacts, requires cooperation and compromise among different parties. It is a pressing challenge for researchers and practitioners to develop and validate novel methods for transportation planning that deliver streamlined planning approaches and improved environmental benefits beyond those possible through traditional approaches (Spellerberg, 1998; Stefanakis & Kavouras, 2002; Mongkut & Saengkhao, 2003; Huang et al., 2003; Gregory et al., 2005). The integration of transportation demand, current and long term development plans, and economic and ecological impacts in time-series scenarios by using land cover and land use analysis is a good way to provide promising results (Saunders et al. 2002; Forman & Alexander, 1998). The use of Multi-Criteria Decision Making (MCDM) as a decision-making framework for transportation infrastructure planning, which can accommodate, model, and combine varying stakeholder values and help to resolve conflicting opinions, is an area that has only been recently explored. Initial results offer significant promise to streamline the National Environmental Policy Act (NEPA) process (Nobrega et al., 2009).

MCDM can facilitate the integration of different planning scenarios as well as the combination of different approaches for environmental sustainability in transportation planning. In modern transportation projects, considerations of both landscape analyses and natural-economic sustainability are mandatory under programs such as NEPA and similar state and local-level laws (Corrales et al., 2000). In 2003, Burnett and Blaschke demonstrated that advances in informatics and geographic information tools have made it possible to segment the complex environments supported by the ecological theory into factors that may be considered in a landscape analysis approach. Current reviews about geospatial landscape analysis in ecology reflect the relatively recent trend towards the use of remote sensing through object-based image analysis (Blaschke et al., 2001; Burnett & Blaschke, 2003;Aplin, 2005). Geographic Object-Based Image Analysis (GEOBIA) employs polygons as bounding areas which delimit the landscape and enable data and image analyses that transcend traditional per-pixel approaches such as spectral-based analysis (Nobrega, 2007; Hay & Castilla, 2008). The use of object-based segments for landscape analysis enable the generation of a large number of parameters based not only on intrinsic values extracted from the polygons, but also extrinsic values computed from the geometry, texture, and context of the objects. This information can be used to form a classification decision hierarchy and provide results that may be combined with existing GIS information to offer significant and innovative results to benefit transportation planning and management and streamline the Environmental Analysis processes (Nobrega, 2007).

BACKGROUND

Watersheds as Natural Biophysical Landscape Segments

Hydrological watersheds are natural subdivisions of the landscape and exercise influence on other natural and man-made features. Wetlands are among the most sensitive of natural features and are vital components of the habitat requiring protection from adverse impacts that may be caused by human development and infrastructure projects. Indeed, NEPA requires transportation planners to consider possible impacts on the hydrological system including stream crossings, flood plains, land cover, and wetlands as part of maintaining the ecological and biophysical balance within the local watersheds (Amekudzi & Meyer, 2006).

This research describes the use of a collaborative, interactive, and iterative multi-scale approach to assess and rank hydrologically segmented features and wetlands to deliver enhanced understanding of how these biophysical systems are affected by transportation infrastructure projects. This chapter addresses a two-level object-based landscape analysis computed from hydrological sub-watersheds from Hydrologic Unit Code 12-level (HUC-12), wetlands, and a subsegment of the proposed Interstate 269 (I-269, a proposed bypass around Memphis, Tennessee, in the southern United States of America) as major objects of interest. Firstly, parameters are extracted per watershed from percentage of wetlands, zoning, existing and current developments, and density of perennial and intermittent streams. Watersheds are ranked according the potential for risk on the natural environment, as described below. The watersheds are considered as primary objects in this hierarchical landscape analysis. After ranking these objects, the next step in the hierarchical analysis process is identifying and ranking wetlands based on potential for adverse impact. For each watershed, topographic analysis (computed from LiDAR elevation data) and computer-assisted image interpretations are performed to enhance the delineation of the wetlands. Wetlands are analyzed according their distance from planned developments, planned roads and the I-269 corridor.

It should be recognized that there are limitations inherent to geospatial data and their analysis within any research framework, and the practical implementation of innovative contributions for geospatial analysis depends upon properly designing and structuring approaches that may be implemented in a practical and feasible framework available in readily available GIS software. In this paper, a top-down GIS framework for landscape analysis is proposed using hydrological watersheds as reference objects for segmentation of the landscape. This segmentation facilitates the geographical analysis of biophysical subdivisions of the landscape based on a watershed approach to conduct contextual, geometrical, and hierarchical analysis. The overall idea is quite similar to standard approaches in object-oriented landscape analysis; however, the use of watersheds as a segmentation layer enables the analysis to consider biophysical subdivision as parts of transportation corridor planning and enables the use of output results in cumulative cost surfaces that may be employed to refine land use and corridor plans and improve agency coordination during the NEPA process.

Landscape Analysis

Landscapes are shaped by the interaction of social and ecological systems (Brunckhorst, 2005). Current and future use of land, productivity and patterns of sustainability are continually modified by humans within the landscape in spatial scales across time in different magnitudes (Onc et al., 2005). For environmentally-focused transportation planning, eco-regions and hydrological watersheds are keys concepts that must be considered in landscape analysis. Understanding landscape and watershed characteristics, the geographic context of sensitive environmental resources, and the services provides by natural systems, is vital to providing balanced solutions for sustainable development amidst natural resources that face economic and social issues (Figure 1). Despite the similarity in some points of view between creating subdivisions of eco-regions and watersheds, a common misunderstanding of each of these landscape subdivision frameworks has resulted in inconsistency in their use and, ultimately, to ineffective application in addressing landscape analysis (Omernik & Bailey, 1997).

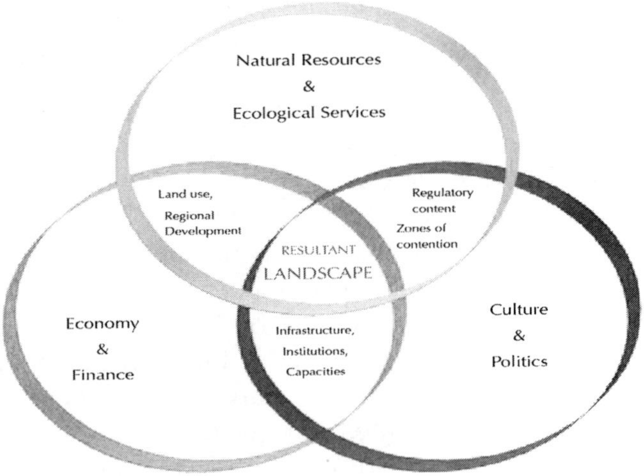

Figure 1: Complex spheres of interaction reflecting human values, identity, and activities affecting landscape change (Brunckhorst, 2005).

Geographic Object-Based Analysis

The traditional methods of classifying remote sensing data are based upon statistical and cluster-based classification of single pixels in a digital image (Lillesand & Kiefer, 2004). Recent research indicates that pixel based classification methods may be less than optimal in producing high-accuracy land use / land cover maps since they do not consider the spatial relationships of landscape features (Schiewe et al., 2001). For example, a significant proportion of the reflectance recorded for a single pixel is derived from the land area immediately surrounding the pixel (Townshend et al., 2000). Analyzing at the polygon object scale enables imagery classification to move beyond this traditional problem.

Contemporary object-based landscape analysis uses parameters derived from hierarchy, context and geometry of the image objects rather than pixel values. Despite a successful history with remote sensing, the accuracy of pixel-based image analysis can be compromised when applied to high resolution images (Nobrega, 2007).

The development and practice of object-based classification has grown as have the variety of methods and approaches of incorporating spatial context into the classification process. Most object-based approaches compliment the axiom of landscape ecology; that it is preferable to work with a meaningful object representing the true spatial pattern rather than a single pixel (Blaschke et al., 2001). Furthermore, the development or use of objects (at one or multiple scales) is always an initial primary phase of the analysis which emphasizes capturing, extracting, or refining the size, shape, and distribution of features of interest.

Object-based classification can be functionally decomposed into two major steps: segmentation and classification. In the segmentation step, relatively homogeneous image objects (polygons) are derived from both spectral and spatial information (Benz et al., 2004). In the classification phase, image objects are labeled as to their class membership by using established classification algorithms, knowledge-based approaches, fuzzy classification membership degrees or a combination of classification methods (Civco et al., 2002).

The commercial software package, Trimble eCognition Developer (formerly *Definiens Developer*), has been well received as a tool for performing object-based classifications of land cover (an example list of scientific papers using eCognition for various land cover mapping tasks is available at*http://www.ecognition.com/learn/resource-center/show-more?type=Scientific%20Paper*). For automated generation of segmentation objects, the application uses a region growing multi-scale segmentation algorithm for the delineation of image objects. The application also enables pre-existing spatial features to be used as objects within which segmentation may be constrained. eCognition provides two different classification methods that may be used separately or combined: a sample-based nearest neighbor classifier with fuzzy logic capabilities and a classifier that enables the development of hierarchic class-membership through a set of rule-based fuzzy logic membership functions.

This chapter presents an implementation of constrained segmentation in which naturally occurring objects provide the initial basis for identifying relevant features on the landscape within which classification and analysis that implement GEOBIA theory are explored.

No segmentation objects are computed, since the objects of interest (watersheds and wetlands) already exist, and segmentation statistics are generated for these areas and used in subsequent phases of analysis. The method combines intrinsic and extrinsic information extracted from the objects and the analyses are organized hierarchically.

Spatial MCDC-AHP in Transportation Planning

Driven by the need to find a balanced solution among conflicting scenarios and because of the vast and growing availability of geospatial data, decision making theory has been explored by the environmental assessment community, including transportation planners.

Multi-Criteria Decision Making is a systematic methodology to generate, rank, compare, and make a selection from multiple conflicting alternatives using disparate data sources and attributes (Gal et al., 1999; Nobrega et al., 2009). The applicability of MCDM is being extended to many different fields including GIS, which is capable of handling massive amounts of geospatial data. Analytical Hierarchy Process is a decision making approach introduced by (Saaty, 1994) based on pair-wise comparisons among criteria and factors in different hierarchical levels. AHP is presented as an effective technique for combining heuristic inputs from stakeholders to achieve a consensus-based decision. The technique allows competing agency expert views as well as stakeholder opinions to be considered quantitatively in a decision making approach (MacFarlane et al., 2008). In keeping with the spirit of NEPA, AHP does not pre-select any specific alternative; it exposes all potential alternatives to the analysis and selection process.

AHP is robust and easily implemented in GIS for geospatial analysis. Results demonstrated in (Sadasivuni et al., 2009) and (Nobrega et al., 2009) showed that AHP can provide significant benefits in facilitating multi-criteria decision-making for planning. AHP is a tool useful for planning and can lead to stakeholder buy-in on planning approaches that consider resource allocation, benefit/cost analysis, the resolution of critical conflicts, and design and optimization. This chapter explores a practical application of spatial MCDM-AHP for transportation planning. The solution presents a semi-automated approach based on an adaptation of Dr. Saaty's theory.

THE STUDY AREA: INITIAL PROCESSING

The Interstate 69 is a proposed 1,600-mile long corridor that connects Canada to Mexico. The entire corridor is divided into 32 Segments of Independent Utility (SIU) for transportation planning and construction purposes. SIU-9 ranges from Millington, TN down to Hernando, MS, crossing the metropolitan area of Memphis, TN and reusing some existing roads such as I-55. However, a new I-269 bypassing the metropolitan Memphis, TN area to the east has been approved through an Environment Impact Statement (EIS) process and is entering the construction phase (Figure 2). The I-269 bypass is the test-bed for a series of research projects sponsored by the National Consortium for Remote Sensing in Transportation -Streamlined Environmental and Planning Process- (NCRST-SEPP). This work is concentrated in Desoto County-MS, which is traversed by the designed I-269.

The NCRST-SEPP project (*http://www.ncrste.msstate.edu/*) applied remote sensing technology and geospatial analysis to streamlining the EIS process for a specific on-the-ground transportation project. NCRST-SEPP research was designed to demonstrate the innovative application of commercial remote sensing and spatial information technologies in specific environmental and planning tasks and activities, validating the use of those technologies by conducting rigorous comparison to traditional methods (Dumas et al., 2009).

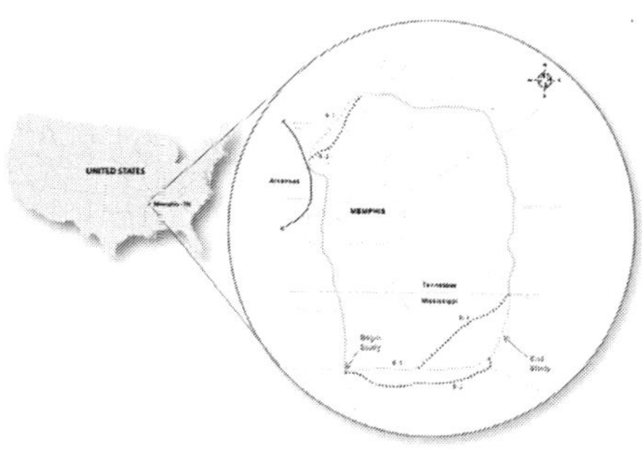

Figure 2: The route of the I-269 bypass including alternatives considered during the EIS process. The study extends along the I-269, in Desoto County, Mississippi, near Memphis-Tennessee.

To make the proposed top-down watershed-wetlands framework analysis useful, this work utilized local geodata provided by Desoto County, MS, such as the transportation network, hydrographical data, LiDAR elevation data, zoning and the county comprehensive plan. A large collection of three-inch resolution aerial images provided support to enhance evaluation of wetland locations. Additionally, wetlands and hydric soil information extracted from satellite radar imagery were used to cover the lack of National Wetlands Inventory federal wetlands data for this specific area (Brooks et al., 2009).

Overcoming the Lack of NWI Information in North Mississippi

In our investigation of efficient methods to provide early assessment to wetlands potentially impacted by transportation corridors, we adopted existing findings of woody wetlands in North Mississippi. According to (Brooks et al., 2009), the motivation in improved methods of mapping forested (or "woody") wetlands areas was two-fold: National Wetlands Inventory (NWI) digital mapping information of wetlands location is unavailable for approximately ¼ of the lower 48 U.S. States, including northwest Mississippi, based on the U.S. Fish and Wildlife Service NWI *"Wetlands Online Mapper"*; and forested wetlands are very poorly mapped using traditional mapping methods including optical remote sensing (Sader et al., 1995; Bourgeau-Chavez et al., 2001).

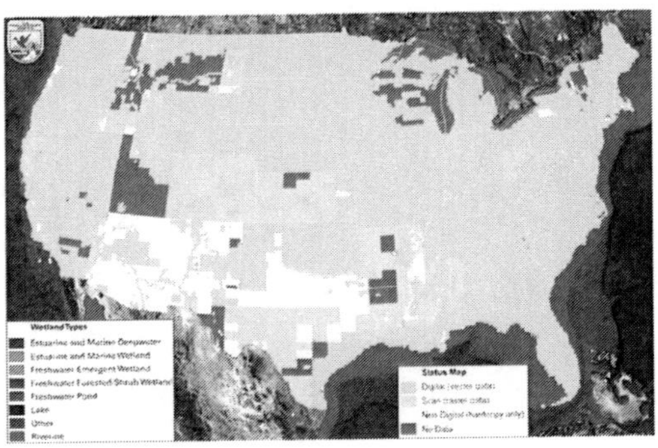

Figure 3: Availability of national wetland data (Modified from U.S. Fish and Wildlife Service, National Wetland Inventory, background image source: Google Earth.

Given this data gap and problems with available traditional sources, we adopted the results described by (Brooks et al., 2009) that used a combination of radar remote sensing data with object-based techniques to compute potential woody wetlands and create a soil moisture index map for the NCRST-SEPP project (Figure 4).

THE TOP-DOWN WATERSHED-BASED LANDSCAPE ANALYSIS

The partition of the landscape into hydrological watersheds was a logical ecologically-focused way to explore the context interactions between the natural and the man-made features. The methodology employed concepts of object-based geographical analysis to evaluate the level of landscape impact of the proposed transportation corridor scenarios. The focus on hydrological watersheds as principal objects made the main difference in comparison with the traditional object-oriented landscape analysis. Two levels of hierarchy were addressed in this work:

- Watersheds were identified and ranked according certain criteria as a significant percentage of unfavourable zoning, density of streams, wetlands and future man-made constructions.
- Wetlands identified and ranked for each watershed. This used topographical LiDAR data, image interpretation and the wetlands impacted by the designed I-269 corridor.

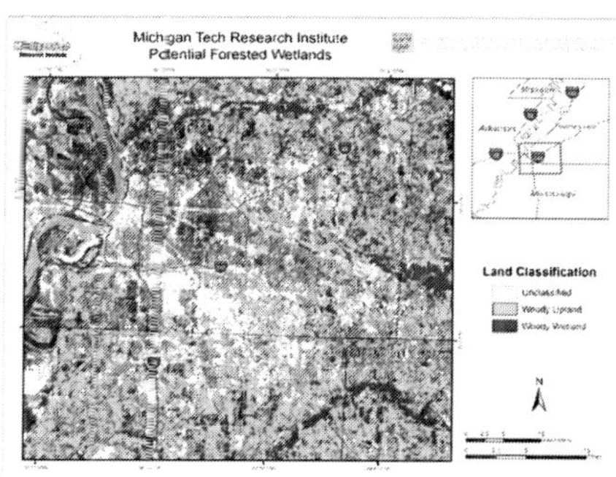

Figure 4: Potential woody wetlands product for the NCRST-SEPP I-269 study area.

Defining Objects in a Hierarchical Landscape Analysis in GIS

An I-269 area GIS was developed to improve the capabilities of geographical analysis by providing ways to access, process, store and disseminate large amounts of information in comparison with human tasks. The traditional GIS features (points, lines and polygons) enable a series of spatial operations as union, overlapping, intersection, etc. Some of these operations were used when integrating the watershed polygons and the landscape layers of information on the first level and when assessing and refining wetlands on the second level (Figure 5).

Level I: Identification of Watersheds

The first step was the identification of the HUC-12 watersheds intersected by the part of the I-269 bypass that included alternative routes in the southeast part of the area located in northwest Mississippi. A simple spatial intersection operation highlighted ten watersheds as shown in Figure 6.

Selecting the watersheds intersected by the I-269 corridor options area caused a significant reduction in the field study area and, consequently, the optimization of the data to be processed. Thus, the next step assessed the numerical criteria that enabled ranking the selected watersheds. At this point, the polygons of the 10 selected watersheds were intersected with other layers of information such as 100 year floodplain, hydrograph, existing roads and urban features, planed roads and developments and zoning in order to extract features to quantify the system. Figure 7 illustrates the layers used in this intersection.

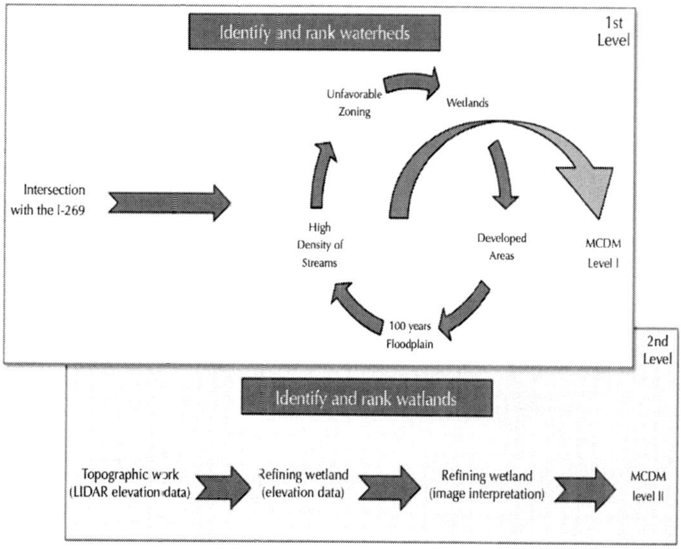

Figure 5: The basic workflow for the two levels of the hierarchical process.

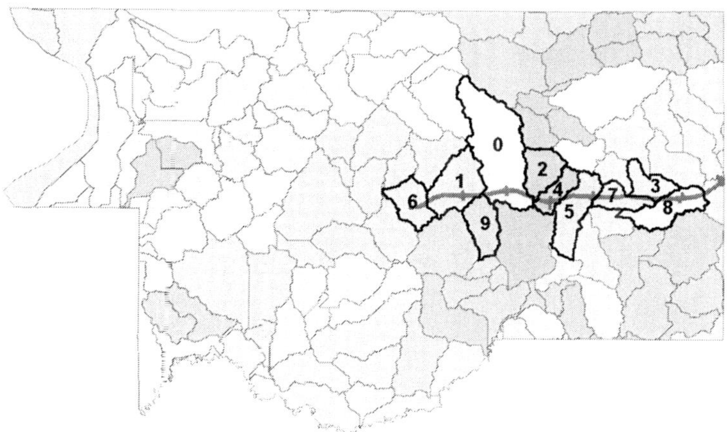

Figure 6: The HUC-12 watersheds intersected by the part I-269 analyzed in Desoto County-MS, which was the section with corridor options assessed as part of the EIS.

In order to make straightforward the process, the zoning map in particular was previously reorganized into 5 classes: agriculture (green), residential (yellow), agriculture-residential (light green), commercial (red) and industrial (orange). Similarly, the maps of existing roads and existing buildings provided by Desoto County GIS Department were combined to produce a density map that reflects the urbanized areas. These steps were necessary since no map of this kind was found to be available in existing GIS databases.

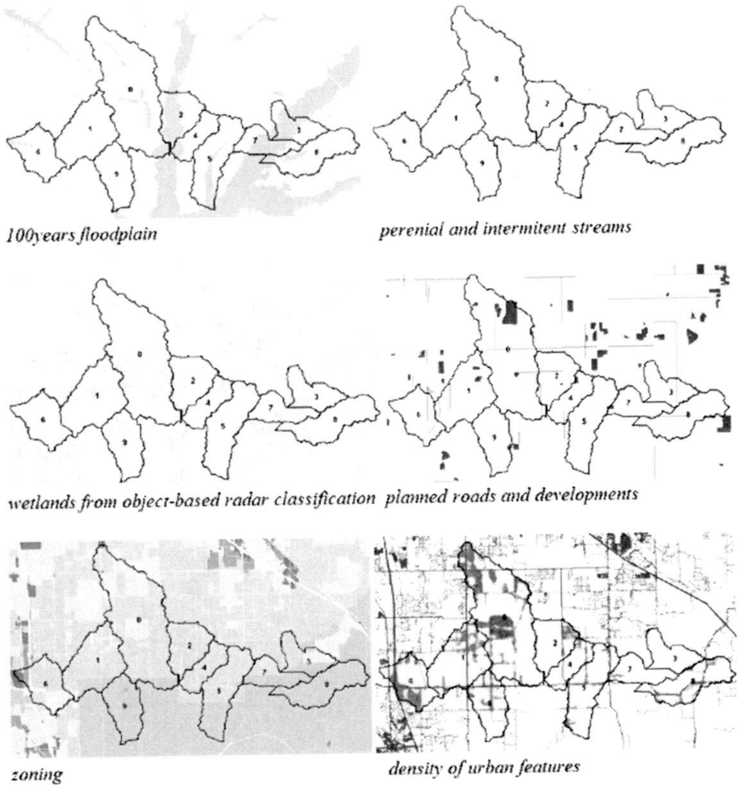

Figure 7: Spatial intersections between the selected wetlands and the layers of interest to assess watershed characteristics to be used in the MCDM process.

Aiming to simplify the decision making process, quite a few different impact factors were assigned to the layers of interest, ranging from 1 (low impact) to 9 (high impact). These values were hypothetical, but reflected the importance of the features due to the potential environmental impact upon existing wetlands. The percentage of covered areas was computed per watershed for the following GIS layers:*watersheds, 100 year floodplain, dense urban, future developments, residential, agriculture, agriculture-residential, commercial and industrial*. Similarly, the density of linear features (km per Km^2) was computed per watershed for the layers *perennial streams, intermittent streams* and *planned roads*. Table 1 presents the relative values extracted per layer from the selected watersheds. These numbers were used to compute the ranking through the weighted average, as show in the Equation 1.

$$\text{Rank} = [\ (A+C+E+F+H) + 3*(J) + 5*(B+I) + 7*(K+L) + 9*(D+G)\ ] / 50 \qquad (1)$$

Table 1: Relative values computed per layer per watershed (percentage of area and density of linear features)

	Layer A	Layer B	Layer C	Layer D	Layer E	Layer F	Layer G	Layer H	Layer I	Layer 1	Layer K	Layer L
Impact factor	1	5	1	9	1	1	9	1	5	3	7	7
watershed	Floodplain %	Dense Urban %	wetlands %	Future Develop. %	Streams km/km2	km/km2	planned roads k/ km2	Agriculture %	Residencial %	Agri-Resid %	Comercial %	Industrial %
0	10.18	35.42	10.28	5.56	0.99	7.96	0.30	5.24	24.08	66.51	0.57	0.07
1	1.58	10.91	10.06	1.75	0.95	8.27	0.25	15.00	10.31	116.31	0.00	0.00
2	65.57	20.24	9.14	0.44	1.79	5.21	0.52	0.00	0.00	128.18	0.00	0.00
3	62.66	3.96	21.09	0.00	0.99	4.64	0.27	33.41	24.21	42.02	0.00	0.00
4	17.08	11.16	5.46	1.50	0.67	6.84	0.16	0.56	0.00	96.59	0.19	0.00
5	18.11	17.03	3.36	0.04	0.93	7.10	0.00	35.53	0.00	62.21	0.13	0.00
6	9.80	43.17	3.00	0.00	0.99	7.26	0.37	0.03	29.40	111.16	3.89	4.30
7	50.85	5.47	23.47	1.39	0.68	3.33	0.14	60.16	0.00	38.57	0.43	0.00
8	11.66	20.24	6.09	0.11	1.01	7.38	0.21	96.51	0.00	0.00	0.04	0.00
9	3.63	28.06	15.05	0.59	0.85	8.79	0.00	95.54	0.00	1.99	0.42	0.00

Level II: Identification of the Wetlands

Unlike the federal and small-scale geodata, local (large-scale or ground-level) geodata normally demand substantial time to be computed due to the high resolution and accuracy involved. Aerial images, high resolution satellite images and LiDAR are the most data intensive information in GIS in terms of storage and interpretation requirements. Minimizing computational efforts by analyzing the landscape, subdividing the geography into semi-homogeneous units, selecting units for further detailed analysis, and prioritizing areas of interest is key to reducing the geographic extent of the study, reducing the computational cost of the study, and supporting the top-down approach in geospatial analysis in which the analysis funnels options down into a reduced set of possible alternatives.

Given the completion of Level I processing described, a series of GIS analyses using information extracted from the topographic surface, such as topographic depressions and flat areas, as well as image analysis such as land cover, provided enhanced inputs to refine wetlands feature geometry as well as classifications based on the radar-based wetland mapping results.

The following processes were developed for the watershed #0, which is second in the ranking as shown in the results section (Figure 8); it serves as a representative example for more detailed examination in this chapter. The reason is that the top-ranked watershed covers a small area and is mostly composed by developed areas and does not present a large variety of landscape features (including wetlands) to illustrate the exercise proposed in this work.

The geospatial analysis was performed using map algebra, so features in vector formats were converted to raster format. Due to landscape analysis considerations and the potential implications of I-269 and planned roads, the layers of information selected in this level (Level II) emphasize the hydrographic and physical aspects, which are basis for engineering construction perspectives.

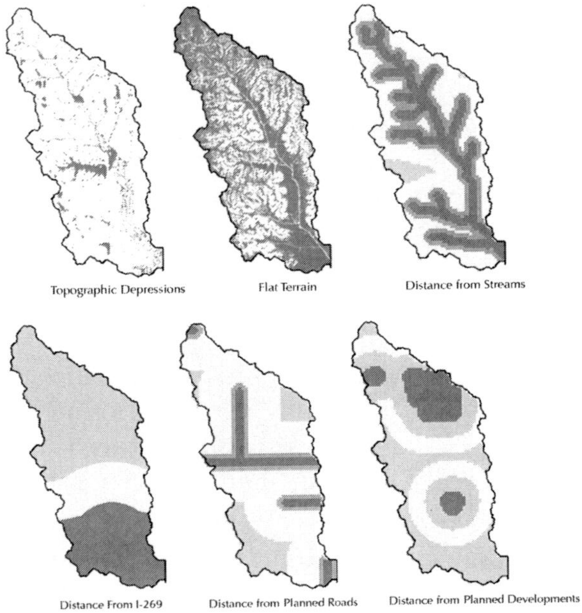

Figure 8: The layers employed to refine and rank the wetlands per watershed. For display purposes, the impact factor ranges from green (low) to red (high).

Table 2 presents the layers used to refine the wetlands features, their respective criteria for classification and the weight to be used on MCDM. Distance criteria and weights are hypothetical; however, they reflect the goal of the paper on assessing potential impacted wetlands.

Table 2: Layers, criteria and weights used on level II analysis

LAYER	**CRITERIA**	**WEIGHT**
Topographic depressions		5
Flat terrain	Slope equal or less than 5%	3
Distance from perennial streams	0-100m, 100-300m, 300-1000m, "/> 1000m	5
Distance from I-269	0-300m, 300-1500m, 1500-3000m, "/> 3000m	9
Distance from planned roads	0-300m, 300-1500m, 1500-3000m, "/> 3000m	7
Distance from planned develop.	0-300m, 300-1000m, 1000-3000m, "/> 3000m	7

The weights are included in the multi-criteria decision tool as input rankings. The tool was developed as part of the SEPP-NCRST project and implemented based on Saaty's AHP method (Figure 9). The normalized weights are used as factors in the map algebra equation that is responsible to produce the cumulative cost surface, where high "cost" would represent higher environmental impact.

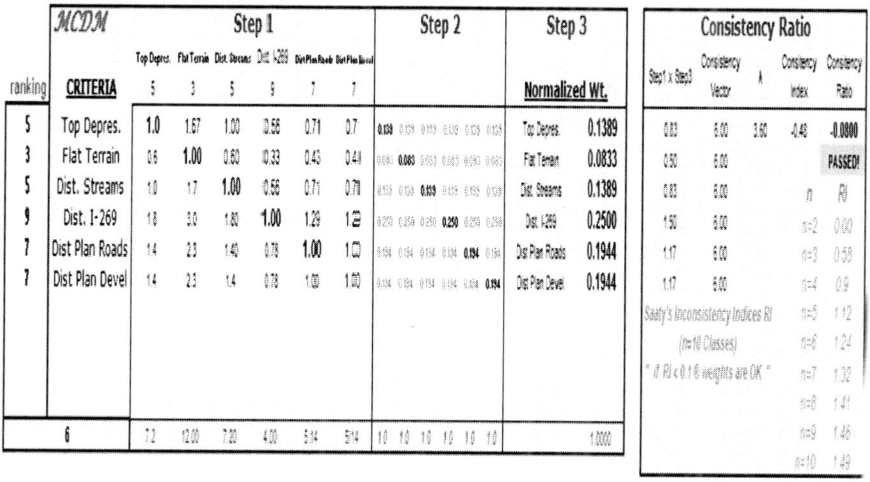

Figure 9: Multi-criteria decision making tool developed to compute normalized weights for the map algebra.

RESULTS

Step 1

For the selected watersheds, impact factors were used to calculate a first-level ranking for watershed and wetland areas impact ranking. Watersheds were ranked and are shown in table 3 in relative order from highest of 15.5 (left-most) to lowest of 4.5 (right-most) on the table.

Table 3: Computed ranking of potentially impacted wetlands intersected by I-269 in the analysis area

Watershed ID	6	0	2	1	4	3	5	7	9	8
Ranking	15.6	11.8	11.5	10.2	7.9	7.8	6.8	6	5.6	4.5

Step 2

For a selected watershed, shown in figure 10 as the watershed #0, the cumulative cost surface shows that impacts are greatest in the lower part of the watershed (Figure 11). In this area, the amount of upland drained is highest; the floodplain is broader and the wetlands are more frequent as are areas of likely ponded water (surface depressions). Over and above the actual number of stream crossings or acres of wetland impacted by a proposed transportation system, this analysis step illustrates that the landscape and hydrologic context of the ecologic and hydrologic features impacted can be shown to play a significant role in assessing the overall impacts of a transportation project on the hydrologic and biophysical systems traversed.

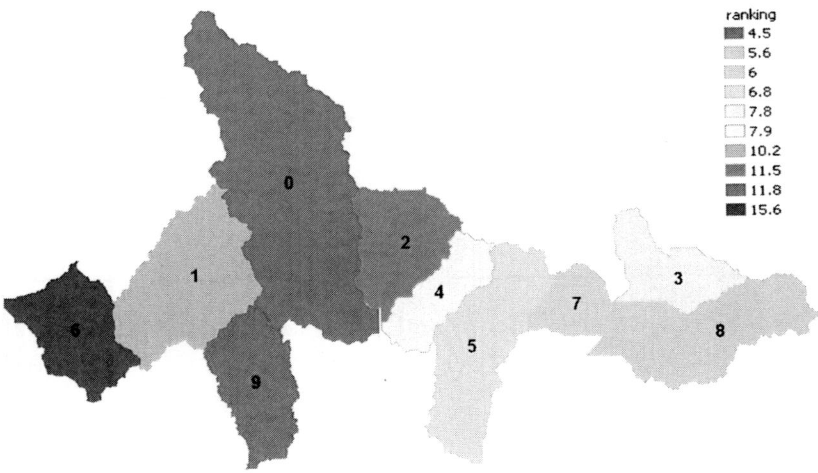

Figure 10: Potentially impacted watersheds intersected by the I-269 – Level I of proposed methodology.

DISCUSSIONS AND CONTRIBUTIONS

In addition to the landscape analysis and transportation planning issues, the results of this investigation showed that a top-down analytical framework based on GIS and MCDM offers value to the early assessment of potentially impacted areas affected by future transportation networks. The work was developed using a set of geospatial data ranging from federal to county, and were intentionally selected to be included in a multi-scale geographic object-based analysis. The hierarchical decision making framework supported the top-down approach through a simple customization of AHP in a GIS environment. The methods and rankings were hypothetically selected (and intentionally made simple) in order to encapsulate the idea and test the MCDM methodology for analysing the impacts of the I-269 study area.

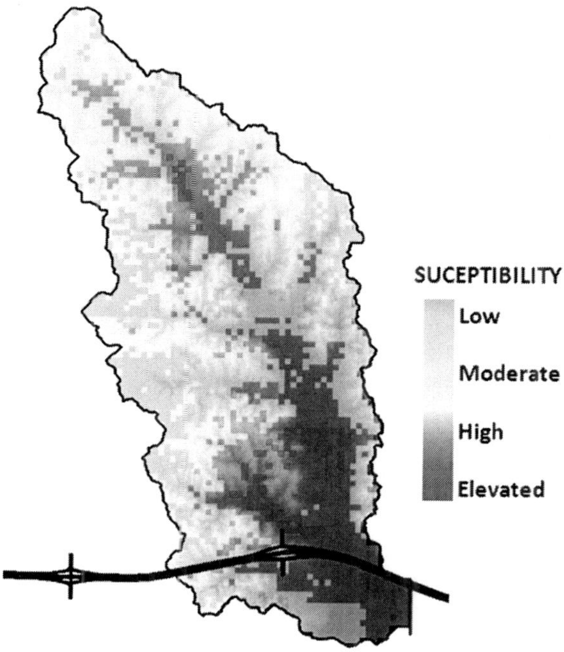

Figure 11: Potential impacted wetland areas (red-orange) highlighted in the cumulative cost surface.

Indeed, in an actual implementation, the relative weights assigned to factors and the rankings associated with factor properties would be subject to alternative assignments of values which would produce results that could significantly depart from those presented and enable a rich evaluation of potential alternatives, including ones based on agency and public inputs. The results presented show a single scenario to illustrate the process rather than an exhaustive exploration of possible scenarios which might arise from collecting a cross-section of objective and subjective values from stakeholders.

The results demonstrate that the proposed top-down approach is a practical screening process valid to the early assessment and ranking of the impacted wetlands.

Desoto County (MS) is an example of many areas that are not fully mapped or adequately covered by Federal mapping efforts such as the detailed county-based soil surveying program and state-based wetland inventories. Therefore, the methodology demonstrates that the complementary use of wetlands computed from radar-based remote sensing can be used to overcome gaps of in the National Wetlands Inventory (NWI) (Bourgeau-Chavez et al., 2009). Indeed, a specific benefit of using remote sensing data is that they can enable the identification of features of interest where ground-based observations or surface-mapped results are limited or absent. For this reason, it is important to highlight that NWI and other relevant data, such as soil survey GIS layers, are not available nation-wide in high detail. Thus, the methodology presented in this paper can be reproduced from environmental and landscape applications, in particular for areas where other map-based products are not available.

The methods presented in this paper were intentionally simplified to highlight a set of framework approaches to help demonstrate the collection of technologies implemented, especially MCDM. Some concepts of geographic object-based analysis such as neighbourhood relationships, contextual analysis, and others could be explored in more depth; however, such an effort would overshadow the desired explanation and synthesis of the more innovative characteristics highlighted in the methodology that focus on MCDM and AHP and their application to context-sensitive landscape analysis in the transportation planning and NEPA processes. Furthermore, it should be noted that the methods presented are both flexible and extensible. They may be adapted to other purposes, transferred to other geographic areas and transportation corridors, and extended to include additional data, steps, and analysis procedures. Follow-on studies are suggested to further explore the application and extension of the methods presented.

CONCLUSIONS

This chapter presents novel methods that leverage spatial implementation of MCDM-AHP in the integrated application of geospatial data to assist transportation decision making throughout the NEPA process. A significant finding is that advanced technologies in geographic object-based analysis can be used to partition the landscape into hydrological watersheds as a basis for context- and object-based analysis. The methodology employed object-based approaches to analyze the landscape and considered a plurality of data layers to derive ranking and weights for understanding the impacts of transportation infrastructure relative to the watershed as a whole as well as to the landscape position of possible transportation alignments within the watershed. The focus on hydrological watersheds as principal objects highlights an important difference in this new and innovative approach as compared to traditional environmental impact analysis.

Watersheds provide subdivisions which are biophysically and ecologically focused, enabling the application of spatial analysis methods which explore the context-sensitive interactions between natural and man-made features in a landscape. The results indicate that the object-based analysis of landscape context and position can provide understanding and insight for assessing transportation corridor impacts on the environment and ecosystems that extend beyond traditional approaches which simply quantify the number of stream crossings and the areas of wetlands impacted. The results show that example hypothetical but reasonable values can be assigned to various landscape features and that these values may be considered in the context of spatially enabled MCDM-AHP. The hierarchical decision making framework implemented through top-down GIS-based analysis enabled the adoption of the segments of the landscape by hydrologic areas. The combination of data, methods and values per level delivers results significant to making decisions, assessing impacts, and designing mitigation strategies that are contextually aligned and indicate an environmentally responsible attitude and sustainable focus for anthropogenic impacts on the environment

ACKNOWLEDGEMENTS

The authors would like to thank the US Department of Transportation Research and Innovative Technology Administration (USDOT-RITA) for funding the research, as well as the Desoto County GIS Department for providing much of the GIS data.

REFERENECES

1. A. Amekudzi, M. M. Meyer, 2006Considering the environment in transportation planning: review of emerging paradigms and practice in the United States. Journal of Urban Planning and Development. 132
2. P. Aplin, 2005Remote sensing: ecology. Progress in Physical Geography. 29
3. U. Benz, P. Hofmann, G. Willhauck, I. Lingenfelder, M. Heynen, 2004Multiresolution, object-oriented fuzzy analysis of remote sensing data for GIS-ready information. ISPRS Journal of Photogrammetry and Remote Sensing. 58
4. T. Blaschke, M. Conradi, S. Lang, 2001Multi-scale image analysis for ecological monitoring of heterogeneous, small structured landscapes. Proceedings of SPIE, Toulouse, 3544
5. T. Blaschke, S. Lang, E. Lorup, J. Strobl, P. Zeil, 2000Object-oriented image processing in an integrated GIS/remote sensing environment and perspectives for environmental applications. In: Cremers, A. and Greve, K. (eds.): Environmental Information for Planning, Politics and the Public. Metropolis, Marburg. 2
6. M. G. Boarnet, S. Chalermpong, 2001New highways, house prices and urban developments: a case study of Toll Roads in Orange County, CA. Housing Policy Debate. 12
7. B. A. M. Bouman, H. G. P. Jansen, R. A. Schipper, A. Nieuwenhuyse, H. Hengsdijk, J. Bouna, 1999A framework for integrated biophysical and economic land use analysis at different scales. Agriculture, Ecosystems and Environment. 75

8. L. L. K. Bourgeau-Chavez, R. B. Riordan, N. Powell, Miller, M. Nowels, 2009Improving wetland characterization with multi-sensor, multi-temporal SAR and optical/infrared data fusion. Advances in Geoscience and Remote Sensing, InTech, G. Jedlovec, ed. 679708

9. L. L. Bourgeau-Chavez, E. S. Kasischke, S. M. Brunzell, J. P. Mudd, K. B. Smith, A. L. Frick, 2001Analysis of space-borne SAR data for wetland mapping in Virginia riparian ecosystems. International Journal of Remote Sensing, 221836653687

10. D. J. Brinckhorst. 2005Integration research for shaping sustainable regional landscapes. Journal of Research Practice, 1n. 2

11. C. Brooks, T. Dajos, L. Bourgeau-Chavez, 2009MTRI forested wetlands and soil moisture: inputs to the multi-criteria decision model. In: NCRST- SEPP: May 2009 Memphis Workshop, 2009, Memphis, TN- USA. National Consortium for Remote Sensing and Transportation Annual Meeting, 2009 http://www.ncrste.msstate.edu/sepp_workshops/memphis_workshop_2009/presentations/mtri_memphis_yr2_rev2.pdf (accessed in July 10, 2010)

12. C. Burnett, T. Blashke, 2003A multi-scale segmentation/object relationship modelling methodology for landscape analysis. Ecological Modelling, 168233249

13. M. Corrales, M. Grant, E. Chan, 2000Indicators of the environmental impacts of transportation: highway, rail, aviation and marine transport. Diane Publishing, 244p.

14. J. Dumas, C. O'Hara, B. Stich, 2009Context sensitive solutions and transportation planning: the detection of land-use, historical and cultural resources, community features and their integration into the planning process. Proceedings of Transportation Research Board, Washington DC, January 2009.

15. R. J. Fletcher Jr, Hutto, L. Richard, 2008Portioning the multi-scale effects of the human activity on the occurrence of riparian forest birds. Landscape Ecology. 23

16. R. T. T. Forman, L. E. Alexander, 2002Roads and their major ecological effects. Annual Review of Ecology and Systematics. 29

17. R. T. T. Forman, D. Sperling, J. A. Bissonette, A. P. Clevenger, C. D. Cutshall, V. H. Dale, L. Fahrig, R. France, C. R. Goldman,

K. Heanue, J. A. Jones, F. J. Swanson, T. Turrentine, T. C. Winter, 2003Road Ecology Science and Solutions. Island Press, Washington, DC.

18. T. Gal, T. J. Stewart, T. Hanne, (eds., 1999Multicriteria decision making: advances in MCDM models, algorithms, theory, and applications. Kluwer Academic Publishers, Norwell, MA. 560 pp.

19. A. Gregory, Todd. S. Kiker, Arun. Bridges, Thomas. P. Varghese, Seager, Linkov. Igor, 2005Application of Multicriteria Decision Analysis in Environmental Decision Making, Integrated Environmental Assessment and Management. 2

20. G. J. Hay, G. Castilla, 2008Geographic Object-Based Image Analysis (GEOBIA): A new name for a new discipline. Lecture Notes in Geoinformation and Cartography. Section 17589

21. J. A. Hilty, C. Brooks, E. Heaton, A. M. Merenlender, 2006Forecasting the effect of land-use change on native and non-native mammalian predator distributions. Biodiversity and Conservation. 15

22. B. Huang, R. L. Cheu, Y. S. Liew, 2003GIS-AHP Model for HAZMAT Routing with Security Considerations. Proceedings of IEEE 6th International Conference on Intelligent Transportation Systems, October 2003, Shanghai, China.

23. I. Keller, Largiardèr, C. R. , 2003Recent habitat fragmentation caused by major roads leds to reduction of gene flow and loss of genetic variability in ground beetles. The Royal Society, 270417423

24. M. Kuitunen, E. Rossi, A. Stenroos, 1998Do highways influence density of land birds? Environmental Management, 22

25. T. M. Lillesand, R. W. Kiefer, 2004Remote Sensing and Image Interpretation (4th ed.), N.Y., John Wiley & Sons.

26. T. A. Litman, 2008Land Use Impacts on Transport: How Land Use Factors Affect Travel Behavior. [online]. Victoria Transportation Policy Institute, 58 p. Available from: http://www.vtpi.org/landtravel.pdfAccessed 15 May 2009]

27. K. Mc Garigal, B. J. Marks, 1995FRAGSTATS: spatial pattern analysis program for quantifying landscape structure. [online]. Forest Science Department, Oregon State University, Corvallis-

OR. Available from: http://www.umass.edu/landeco/pubs/Fragstats.pdf [Accessed 15May 2009

28. P. L. Mokhtarian, X. Cao, 2008Examining the impacts of residential self-selection or travel behavior: A focus on methodologies. Transportation Research B, 42
29. R. A. A. Nobrega, 2007Deteccao da malha viaria na periferia urbana de Sao Paulo utilizando imagens de alta resolucao espacial e classificação orientada a objetos. Tese de Dourotado, Escola Politecnica da Universidade de Sao Paulo, Sao Paulo. 165p.
30. J. M. Omernik, R. G. Bailey, 1997Distinguishing between watershed and ecoregions. Journal of the American Water Association. 33
31. S. Ono, M. T. L. Barros, G. Conrado, 2005A Utilizacao de SIG no planejamento e Gestao de Bacias Urbanas. Proceedings of AbrhSIG. Sao Paulo, 2005.
32. T. L. Saaty, 1995Transport planning with multiple criteria: the analytic hierarchy process applications and progress review. Journal of Advanced Transportation. 29
33. R. Sadasivuni, C. G. O'Hara, R. Nobrega, J. Dumas, 2009A transportation corridor case study for multi-criteria decision analysis. Proceedings of American Society of Photogrammetry and Remote Sensing Annual Conference, March 11-14, 2009, Baltimore-MD.
34. S. Sader, D. , W. Ahl, 1995Accuracy of Landsat-TM and GIS rule-based methods for forest wetland classification in Maine. Remote Sensing of Environment. 53
35. S. C. Saunders, M. R. Mislivets, J. Chen, D. T. Cleland, 2002Effects of roads on landscape structure within nested ecological units on the Northern Great Lakes Region, USA. Biological Conservation. 103
36. J. Schiewe, L. Tufte, M. Ehlers, 2001Potential and problems of multi-scale segmentation methods in remote sensing. GIS-Zeitschrift für Geoinformationssysteme. 6
37. I. F. Spellerberg, 1998Ecological effects of roads and traffic: a literature review. Global Ecology and Biogeography Letters. 7

38. Walker. J. Townshend, R., G. Huang, Fries. N. V. Kalluri De, S. Liang, 2000Beware of per-pixel characterisation of land cover. International Journal of Remote Sensing. 21

39. A. P. Wheeler, P. Angermeier, A. Rosenberguer, 2005Impacts of new highways and subsequent landscape urbanization on stream habitat and biota. Reviews in Fisheries and Science. 13

Chapter 3

Feng-Shui Theory and Practice Investigated by Spatial Regression Modeling

Jung-Sup Um

[1]Kyungpook National University, South Korea

INTRODUCTION

Feng-Shui (meaning wind and water) has a strong indication for the placement of a physical object in space since it determines propitious locations by investigating the vital energy of the earth (Q_i) throughout the lay of the land. Feng-Shui is the product of ancient Chinese cosmological beliefs and their responses to the land which were directly linked to survival and healthy living since it was believed that it could be used to harmonize people with their environment. Feng-Shui practices could be divided into landform Feng-Shui and living-space Feng-Shui (Sung, 2001). Landform Feng-Shui practices have been conducted with emphasis on grave, residential sites, village, town, and city sites while living-space Feng-Shui focused on the physical layout of such elements as the front gate, bedroom and kitchen. Today, living-space Feng-Shui is being widely used in the United States and Western Europe (Kim, 2003). In some respects, these two phenomena are quite different. However, they both involve patterns in siting that seem to express cosmological beliefs, numinous feelings and aesthetic experience.

Feng-Shui has been practiced in China and many other areas of Asia over the past five thousand years and it is becoming increasingly popular in western countries, such as the United States and Western Europe, thus gaining popularity world-wide (Chris, 2002; Diana 2004). Feng-shui theory and practice were based on considerable anecdotal evidence rather than on scientific proof. It should be noted that conflicts and debates exist when applying Feng-Shui rules because of the flexible practices and ambiguous literature (Andrew, 1968; Zetlin, 1995). Even Asians, who are familiar with ancient oriental literature and philosophy, have difficulty understanding the basic Feng-Shui concepts and methods of application.

So far the spatial variation of Feng-Shui locations is poorly defined despite abundant interest in this problem. Previous research for living space Feng-Shui has been based on the small number of in-situ observations reflecting the specific attributes of experimental sites such as the window, mirror and bed room (Hobson, 1994; Verlyn, 1995; Lu & Jones, 2000). Field practices for living-space Feng-Shui have been researched with a focus on questionnaires (June, 2002). Field observations have the disadvantage that they provide only limited information on the area-wide distribution of the Feng-Shui locations.

Although area-wide data such as a topographic map for villages, towns and city sites in landform Feng-Shui have been used, it is rare to find enough experimental locations to prioritize the relative importance of spatial variables for Feng-Shui practice in those applications since just one sample used to occupy a huge area (e.g city site) (Hammond,

1995,; Xu, 1998; Zitao, 2000; Jun, 2003; Lynch, 2003; Mak & Ng, 2005). Spatially prioritized parameters cannot, therefore, be adequately identified on the basis of such a few Feng-Shui locations. Until recently, the investigations of an area-wide distribution pattern of Feng-Shui locations and their spatial characteristics remained largely theoretical because the non-GIS survey technique had difficulty in assembling multi-thematic factors simultaneously (Earle, 1992). In the case of burial mounds, there is area-wide spatial evidence to document them. The actual grave distribution is a result of the complex interaction of historic and recent environmental and human factors. Tomb footprint shows a tenacious vision of what constitutes a proper mound location. The footprint of a grave characterizes Feng-Shui's spatial structure that varies with a multitude of factors such as the land use pattern and surface roughness. However, the field practices of tomb Feng-Shui have seldom been empirically examined in terms of area-wide spatial arrangements. It is necessary for the complicated belief system of Feng-Shui to be explained by describing how tomb footprint is distributed in space or with respect to each other. In Feng-Shui, every piece of information has to be looked at individually and put it together with others and look at the whole (or bigger) picture. Everything has its own place in every different situation and position. Concentrating on one point to the exclusion of all others does not lead to a balanced point of view. GIS is an ideal technology capable of integrating, merging and analyzing simultaneously multiple data layers to explore spatially prioritized relationships that are necessary when studying Feng-Shui footprints.

The multivariate spatial regression modeling compares multiple maps at one time and answers questions such as "how much more important is one map over the other". It is expected to reveal a consensus on how the Feng-Shui practice may have a universal sense of what is the most important variable for any place. If the importance or preference of each factor relative to other factors is identified in the context of a spatial multi-criteria framework, high priority areas in Feng-Shui will be identified. Therefore, the aim of this research was to evaluate major controlling factors in locating Feng-Shui tombs, using the spatial prioritization modelling techniques of GIS, based on the area-wide grave footprints in South Korea.

GENERAL METHODOLOGY

Model Parameters

Feng-Shui, by studying the configuration of mountains and rivers, selects a location where the vital energy (Q_i) that flows throughout the earth is connected by water and not scattered by the wind. There is often a lack of understanding as to where this Q_i is coming from. An important question is how to measure or represent criteria that are important to Feng-Shui practices and should be included in a spatial regression model. Preference variables perceived according to different cultures and geographical differences vary significantly across individuals and across the groups these individuals represent. Each grave location in a cluster, or isolated, takes into consideration the astrological conditions at the site for its unique occupant. It is important that the occupants of the sites are rich or poor, or what clans they represent. Facing a mountain at great distance is also considered as one of the important variables of a propitious location. The grave density is subject to complex, interacting influences, several of which, notably competition for favorable landscape and micro-site variability.

It is impossible to measure and quantify many of them for the purpose of spatial modeling since they need to be measured based on people's perceptions and feelings. It would clearly be impossible to model the behavior of every subjective variable involved in the burial decision. This method is extremely subjective since the model has to be run under vast numbers of combinations of rules; its approach is conceptually non-spatial. In such circumstances, there is no objective means of ascertaining the data quality and extent of the model. Such variables are too much subjective to be of great use in the modeling exercise.

Landform is frequently recognised as the most important factor in the grave location since it includes indicative information such as water features, vegetation, landslide susceptibility and soil moisture, etc. for the site investigated (Choi, 2001; Yoon 2006). Water moves through and over the landform, so the spatial distribution of soil and vegetation properties should be related to topographic attributes such as slope, elevation, and aspect and solar radiation in a specific grave location.

The arrangement of meaningful Feng-Shui landscape "mountain in the back, water in the front" is closely related to landform factors such as the direction of maximum slope and transverse to the slope. In this regard, landform is believed to characterize changes in the flow of energy direction (Qi), flow velocity and Qi transport processes based on the orientation of mountains and spatial arrangements of topographic attributes such as slope.

In general, models of physical systems are forced to adopt aggregate approaches because of the enormous number of individual objects involved. It was necessary to focus on the variables which can be quantified to ensure the quality of the model, compared to the traditional use of subjective rules of thumb. These studies have employed only the most readily available and accessible data that are directly integrated with GIS to focus on representative and typical features in the experimental design and in order to evaluate the field Fung-Shui practices in an objective manner.

The main focus has been given to demonstrating the influence of the hillshade, elevation, slope and aspect as the most important factors controlling the spatial location of a grave. Landform multi-criteria models may provide an equitable and efficient means for exploring the spatial structure of Feng-Shui in terms of only stable physical geography. However, new grave location procedures could be influenced by various human factors besides landform parameters. In order for these tools to be effective, they should include criteria that are locally relevant and measurable in a human framework. To examine land-use effects and proximity to population on grave densities, several proximity variables calculated in accordance with the distance from roads were included as the model parameters (Table 2).

Study Area and Data Analysis

Feng-Shui is responsible for many aspects of the cultural landscape in Korea, especially the location of tombs. The process of tomb location is extremely important in South Korea, since it is a common belief accepted by the Korean public that the location of the tomb would affect the descendant's well-being of the tomb's occupant. It is well known that a strong belief in Feng-Shui led kings in past dynasties to exhume and rebury their ancestors. "Even former and current presidents, and leaders of major political parties in South Korea, have moved their ancestors' graves or sought out propitious locations. Also, it was thought that burying one's parents in a place deemed propitious under Feng-Shui principles was a way to demonstrate filial piety (Kim, 2002)." The direction and position of grave were carefully selected by geomancers. "The spirit of Feng-Shui is still visible and strongly felt in the countryside, tightly integrated into the rural way of life" (Choi, 2001).

The burial areas are well maintained by a clan and memorial ceremonies are conducted on the traditional ancestor worship days such as Thanksgiving Day and New Year's Day, etc. at the tomb (Hough 1999). In this process, "Feng-Shui plays the role of a chain of life that connects an individual with his ancestors as well as descendants" (Yoon, 2006). "This is why Feng-Shui has continued to be influential despite being constantly criticized as a superstition (Kim 2002)". Such practices make the Korea the best place to explore the principles of Feng-Shui in the entire world.

The study area is situated in the south eastern part of South Korea between latitude 35.55°N and 36°N and longitude 128.38°E and 128.44°E. It belongs to the Gongsan-Dong (administrative districts) in the city of Daegu, a metropolitan city recognized as the third most important city in South Korea, covering approximately 56.5 km2 (Fig. 1). It possesses several advantages that make it an appropriate choice for such a study. Its elevation ranges from approximately 60–834 m above sea level and shows a hilly character, frequently observed as the topography of typical grave locations in South Korea (1). The area is located on the south eastern edge of Palgong Mountain which is well known for geomantic ideals as well as one of the most spiritual mountains in the country. In Korea, Buddist monks appear as geomancer-monks who choose most of the auspicious places and thus the locations of many Buddhist temples have geomantic legends (Yoon, 1975). The study area was located in a place where it is easy to find many old temples in every valley and Buddhist heritages are scattered (Korean Heritage Online, 2006). The area is relatively isolated from the urban center, thus grave locations in the area still keep the appearance of the old dynasty period. The study area is located at the place where a cemetery is not available to investigate the natural location pattern of tombs by individual clans.

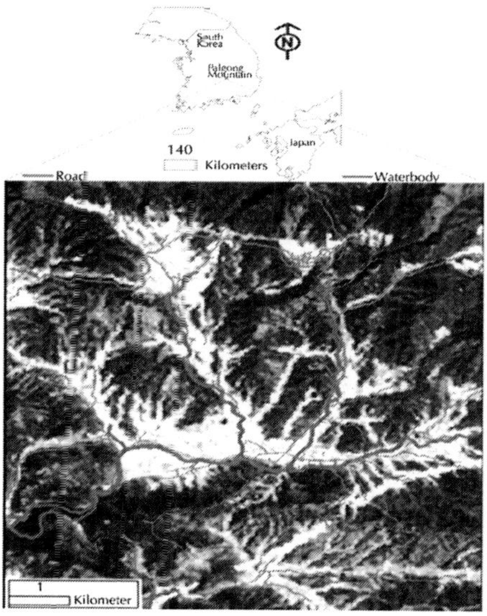

Figure 1: Location map of experimental site presented by a Thematic Mapper satellite image (acquired in April 2002).

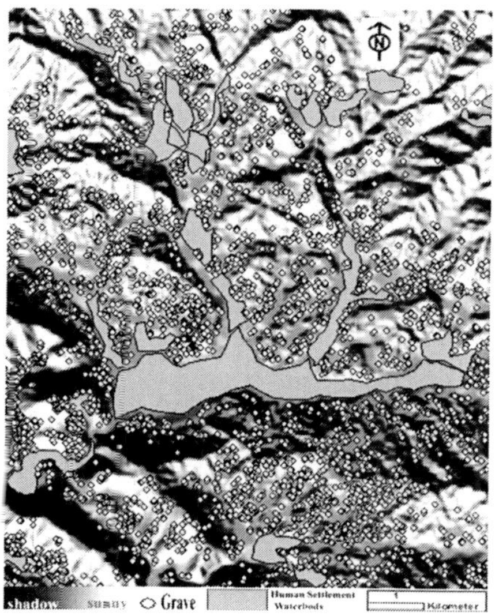

Figure 2: Solar radiation map combined with grave locations. A shadow area at the south eastern end and northern part of the study area shows a very small number of graves.

A1:5000 digital topographic map (produced in 1996 from aerial photography taken during the same year) was accessed via the national mapping authority of South Korea. The base map has been used to extract a Feng-Shui database such as the contour with 25 m equidistance, administrative boundary and road and tomb available in the study site. The human settlements, golf course and water-body have been excluded for further analysis since graves could not be located there (Fig. 2). The analysis in the ArcGIS was performed to identify and calculate spatial variables to be used in a regression model. A localized grave density map was created by defining the mean number of points per unit area in the ArcGIS to pinpoint spatially intensified distribution areas and examine the surrounding causal factors such as landforms. The search radius for the calculation of each pixel (30 m) was set to 300 m by an initial option suggested by the density slicing function of ArcGIS. The distributions of the concentration levels and predictor variables are presented by percentile scale as shown in the Fig. 2.

From the contours and spot heights, a DEM (Digital Elevation Model) was generated with pixel size 30 m × 30 m. Hillshade, slope and aspect layers were created as derivatives of the DEM. Computing the hillshade requires one to enter values that identify the sun›s height and location in the sky. The hillshade grids were produced with an azimuth of 195 degrees, an altitude of 60 degrees based on the Korean holiday (Hansik) falling on the 105th day after the winter solstice (13:05 April 6, 1996) of the study site. It is a day for the family to move their ancestors› tombs and to hold their own memorial services. Each hillshade cell contains a value which indicates the sun intensity, ranging from 1 to 255. It was used to determine what was in shadows and what was is in sunlight at that time in 1996.

The road-related spatial database, population density and land prices were already available for most areas of Korea. The population register (1996) was acquired from the Korean central statistical Office and land prices (1996) from the Korean Association of Property Appraisers (1996). The population and land price statistics were added to each administrative unit by means of joining the data table to a theme's attribute table. Feng-Shui databases were built as integrated GIS layers and were overlaid accordingly. The data acquired from the spatial analysis were stored in MS Excel and basic statistical parameters were calculated to acquire the overall feature of the data sets. Multivariate spatial regression was undertaken by the SPSS statistical package to examine the relationships between grave concentrations and spatial variables potentially associated with grave density (Tables 2). The fit of the spatial models was assessed by the percentage of explained variation (R^2); slope estimates of the parameters and their VIF (Variance Inflation Factor) were given.

RESULT

Summary data for the grave density in relation to proximity to roads is presented in Table 1. The grave density data does not show a significant degree of variability for different proximity conditions from roads. The nearby places to the road are not more related than distant places in terms of grave concentration. Similarly the results of the regression analyses indicate that proximity to a road is not an important predictor of grave density. The population density did not make a significant contribution to explaining the variance in R^2 since the owners of tombs are mostly residents of urbanized areas far away from the study area. The proximity or human variables were the least significant variables in the regression model. Furthermore, when the human variables were excluded, the R^2 values for the models decreased by just 0.009 (0.754 to 0.745), showing the lower importance of road, population, and land price variables. By contrast, grave concentrations in landform parameters showed relatively high variability (Table 2).

Table 1: Area-wide grave density (number of graves per km^2) in relation to buffer distance from road

Road (6 m width)	density	Road (4m width)	density	Road (2 m width)	density
50 (meter)	84	50 (meter)	83.9	50 (meter)	78.9
100	108.9	100	109.4	100	78.7
200	118.8	200	124.6	200	76.6
400	110.4	400	104.1	400	80.2
600	83.6	600	102.4		
1000	98.9	1000	96.5		

Table 2: Multiple regression results including human variables and landform variables alone (dependent: grave density)

	Model 1 (with proximity and human variables)			Model II (with landform variables alone)		
Summary of model	Adjusted R2=0.754, number of observations (65523), F=16719.154 (P= 0.000)			Adjusted R2=0.745, number of observations (65523), F= 27405.752 (P= 0.000)		
Explanatory variables	Standardized coefficients	t-value	VIF (Significance levels*)	Standardized coefficients	t-value	VIF (Significance levels*)
solar	1.590	178.293	21.173	1.557	289.977	7.417
elevation	-.858	-148.941	8.832z	-.806	-152.509	7.189
slope	-.191	-34.111	8.329	-.116	-22.696	6.771
south	.049	18.413	1.894	.054	20.469	1.182
east	.063	23.922	1.818	.069	26.662	1.746
west	.018	8.135	1.275	.015	6.595	1.262
north	.001	.472	1.101(.637)	-.002	-1.145	1.09 (.252)
road (6 m)	.114	33.563	3.086			
road (4 m)	.098	28.204	3.242			
road (2 m)	-.003	-.817	2.776(.414)			
population	-.006	.609	26.13 (.543)			
land price	-.080	-8.695	22.36			

The R^2 values for the models indicate that 74.5% of the variability of grave densities observed in the site was explained by the landform variables (Table 2). Initial observation of the grave map overlaid on the solar radiation clearly disclosed strong evidence for the influence of the sun index as a derivative of the integrated landform parameters at the grave distribution, as shown in Figs. 2-3. Solar illumination made the most significant contribution (b: 1.557) to explaining the variance in grave presence or absence (Fig. 2-3). Solar radiation is affected by various factors such as the latitude of the site investigated, slope and aspect of a site, sun's height and location in the sky. There was a more distinct preference for low altitude sites (Figs. 4-5) and a stronger association (b: -0.806) with decreasing elevation than those of slope (b: -0.116) and aspect. The relationship between the slope and the grave density variable is less distinct and a clear trend cannot be observed. A scatter plot shows that graves occur more frequently where the slope is intermediate (Fig. 6).

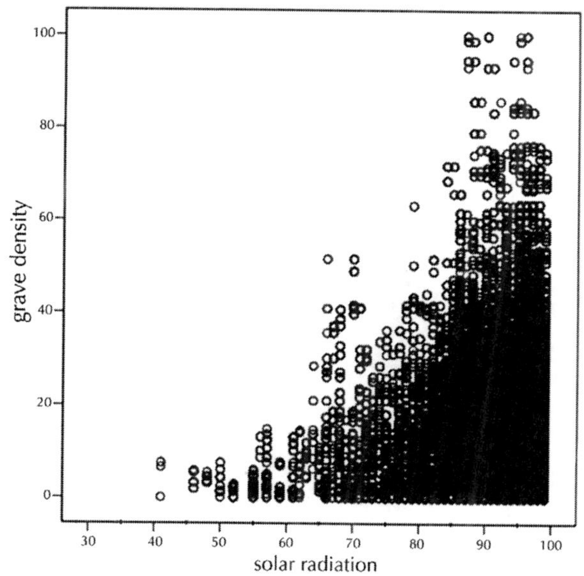

Figure 3: Plot of grave density vs. solar radiation (percentile scale, One hundred percent represents the highest grave density in the experimental site). Local grave density displays distinctively increasing trends in respect to the occurrence of a sunny site condition. The plot shows that the sun index is a crucial control factor on grave density.

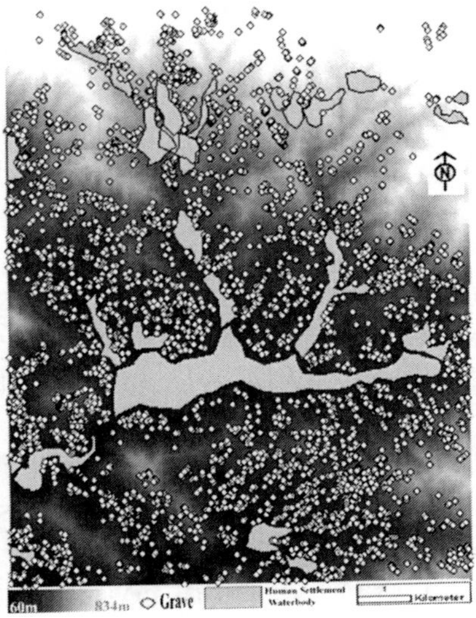

Figure 4: Digital Elevation Model combined with grave locations. A highly elevated landform condition at the northern edge of the study area shows a very small number of graves.

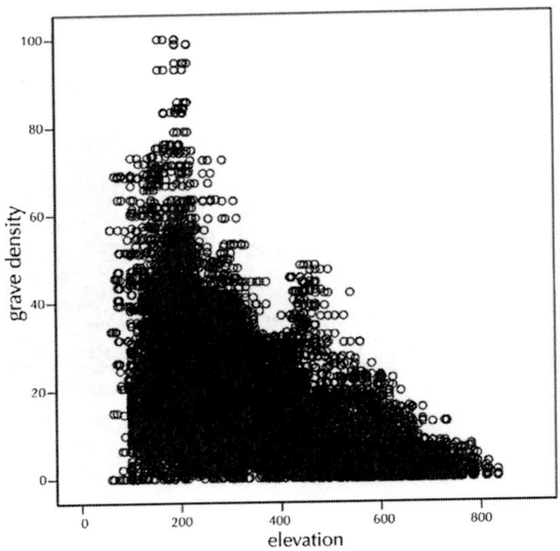

Figure 5: Plot of grave density (percentile scale) vs. elevation (m). Local grave density displays decreasing trends in respect to the occurrence of a high altitude condition.

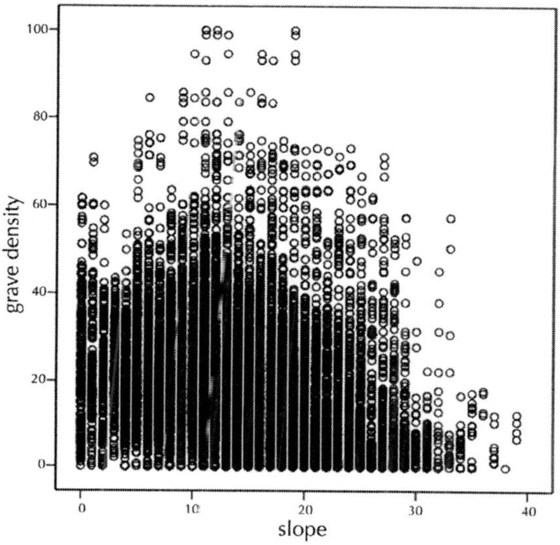

Figure 6: Plot of grave density (percentile scale) vs. slope (º). There is considerable scattering in the individual data points, but a higher grave density is generally found on the intermediate slope range.

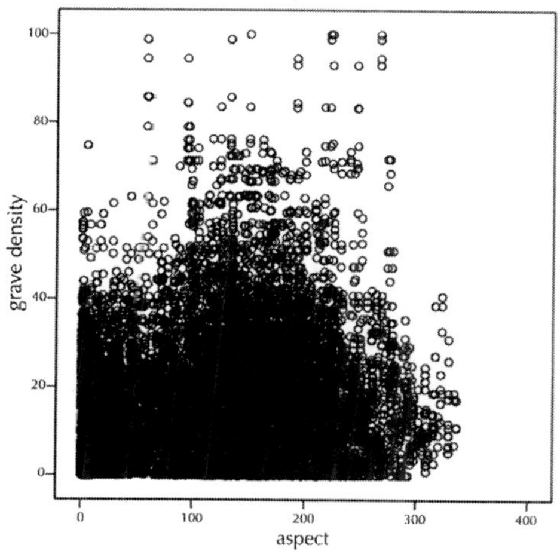

Figure 7: Plot of grave density (percentile scale) vs. aspect (º). Grave density is generally higher on south (136º-225º) and east-facing (46º-135º) slopes than on north (0º-45º, 316º-360º) and west-facing slopes (226º-315º).A distinct preference for south direction is not observed.

Grave density was generally higher on south and east-facing slopes than on north, and west-facing slopes (Table 2). The number of graves in the southern direction is not significantly higher than that of the east or west, different from the common belief that a south-facing is most favorable. The aspect parameter showed a much weaker influence (Fig. 7) on the grave distribution as a whole in the regression model than those of the sun's index (b: 1.557) and elevation (b: -0.806).

A comparison of scatter plots for the sun index with other landform variables indicated that grave density is much more abruptly decreased in a shadow condition than those with other explanatory variables such as elevation or slope. The sudden drop of grave density in accordance with the sun index bands indicates that the grave density is strongly associated to the site condition within the sun index 92 percentile. There were high joint effects for several variables, reflecting the colinearity between the variables. The most significant co-linearity was among these three, the sun index 99 percentile, elevation 251 m and elevation 423 m. Despite this colinearity, these variables were retained in the models because they could display their differing relative importance in the regression trees.

DISCUSSION

Several studies have shown that Feng-Shui locations tend to be dependent on landform parameters, and increase if favorable conditions are observed, like when a site is located in low elevated hills, has a smooth slope and a good orientation facing south (Andrew, 1968; Yoon, 2006). The results of this study extend previous findings, specifically focused on the associations with the landform surface. The present study has utilized area-wide and pertinent information in the analysis of a Feng-Shui location to identify the spatially prioritized variables. In this project, the number of samples (5549) was large enough to extract reliable information relating to the existing state of the grave density in question, as substantial evidence of the spatially prioritized relationship caused by the Fung-Shui application. The area-wide grave data acquired from the topographic map produced by the interpretation of aerial photography provided sufficiently reliable information by replacing the questionable accuracy of a conventional field survey by a permanent record. This study has also demonstrated that a Geographic user friendly interface can be utilized for a factual assessment for Feng-Shui distribution by overcoming any serious subjective judgment suffered from the lack of a cartographic representation in the text based statistical data.

The spatial regression analysis allowed for assessing the level of consistency among the spatial variables in Feng-Shui practices by evaluating their stratifying hierarchy simultaneously. The grave locations were rarely affected by their degree of isolation or their degree of connectivity to roads, the dynamics of local populations and the price of the land segments. The results show that grave locations generally are determined only by the ideal Feng-Shui model established more than two thousand years ago without considering the surrounding environmental changes caused by urbanization. The spatial prioritization model clearly demonstrated that the sun index is perceived as extremely important while the elevation is considered moderately important and the aspect and slope are less important.

The actual importance (regression coefficient) value of the sun index presented by the spatial prioritization model can answer such a question as "how much more important is the sun index over the others?" In usual site selection cases, selecting one site over the other requires a tradeoff among spatial variables, provided the variables have a similar importance on site selection. The key issue in the spatial prioritization model would be a further tradeoff among first, the sun index, then the second most important and a combination of the rest. The magnitude of solar radiation and its relative importance is great enough to offset the difference in other landform parameters. It is believed that the Feng-Shui experts used to take the most important factor, the sun index and make tradeoffs between the second important factor, elevation and the reduced combination set of the last two factors (slope and aspect).

Feng-Shui emerged in the context of agricultural societies located in a temperate zone. It was essential to find a sunny location for farming and to survive in a chilly climate. In this regard, the relative merits of other landform factors in the overall Feng-Shui siting were very weak in comparison with the solar radiation in the spatial model. The sun index is much more influential a factor in the model's predictions since it is jointly determined by various landform parameters and the solar system in different locations. The findings of this study have important implications for the application of Feng-Shui and for future research. Also, the findings of this paper may assist in the more accurate targeting of Feng-Shui applications at their locations. It is believed that the postulates of the classical Feng-Shui theory are applicable to a specific sun light deficient area and not to the general case. In a tropical area, the sun can be quite intensive. It is not necessary for the sunlight to be shining into the sites all day (Tan, 2006). That is the reason why it is so hard to find real evidence of Feng-Shui applications in tropical regions such as South America and Africa.

This research has been carried out in one of the best country to explore the principles of Feng-Shui in the entire world. The experimental site is located on the site well known for geomantic ideals as well as one of the most spiritual mountains in the country. The spatial prioritization model derived in the study area may reflect Feng-Shui practices under ideal circumstances. The spatial dimensions of Feng-Shui locations vary according to the way different societies construct different shared cultural spaces. Moreover, the characteristics of the special case assumed by the classical Feng-Shui theory happen not to be those of the economic society in which we actually live. The perceptions about the surrounding environment generally vary with geographical differences. However, it is expected that the result of a Feng-Shui application will almost always parallel with the modern solar siting techniques and theories as the outcomes of these inspections have illustrated if the ancient Feng-Shui principles have been applied properly. Therefore the spatial prioritization model of this research will provide a standardized form against which to compare the real phenomenon of Feng-Shui. Although the cultural and geographical differences between the Western and Eastern Feng-Shui rarely behave this way, it is still useful to know what would happen if they did, as a basis for comparison.

CONCLUSIONS

Analyzing spatially prioritized parameters in relation to the Feng-Shui surface represents a critical step toward understanding how a local condition contributes to Feng-Shui landscape and will help to evaluate Feng-Shui locating strategies. This is one of a few studies that have looked at the relative importance of the differences of spatial variables within Feng-Shui locations and the first to formalize a solar illumination priority in a Feng-Shui tomb with area-wide empirical evidence. The spatial prioritization model showed that it could play a key role in evaluating Feng-Shui tomb distribution in which different kinds of landform mosaic mixes could be accommodated in the same location and could also occur in different locations. Therefore, this study made a valuable first step toward reducing its aura of superstitious mystery by investigating a major contribution factor for Feng-Shui locations through more rigorous and systematic approaches.

Solar radiation was the most important predictor of grave density in the Feng-Shui locations. Similarly, spatial clustering technology identified the fact that high concentrations of grave necessarily accompany the significantly increasing trends of solar radiation. The results of the regression analyses indicate that the grave density could be explained by the four landform parameters alone yielding R^2 values of 0.751. In contrast to the typical theory, slope and aspect were not a dominant determining factor upon the dependent variable of grave density. Also, the significantly increasing trends of grave density were not observed in line with a southern direction. A clear verification has been made for the hidden assumptions in Feng-Shui's long history that its approach is found to be more appropriate in avoiding shadow conditions, rather than exploring the ideal landform location

However, this research endeavor leaves many questions for further discussion and investigation. This study only finds statistically significant association between grave density and certain variables describing landscape and it does not attempt to elucidate what this association means in cultural terms with a more structured code of practice for Feng Shui application. One possibility is that dispersed patterns of graves could be the typical result of competition for a particularly auspicious piece of land. The presented approach to identify spatially prioritized parameters of Feng-Shui has yet to be tested in more diverse situations (agriculturally, environmentally and socio-economically) and the parameters of the analysis have to be more thoroughly validated through further research. This type of information generally requires extensive cross-cultural and comparative studies. There is a need to compare a variety of cultures equally from an objective viewpoint, to explore the spatially integrative relationship between the grave density and their surrounding landscape. This work constitutes an early step in the process of identification and characterization of the full suite of spatially prioritized parameters of Feng-Shui.

ACKNOWLEDGEMENTS

Thanks are extended to the different national agencies of South Korea (Spatial Data Warehouse of National Geographic Information Institute, Central Statistical Office, and Korean Association of Property Appraisers) for sharing their data for this project. I am grateful to Prof. Michael Goodchild of the National Science Foundation-funded National Center for Geographic Information and Analysis (NCGIA) at the University of California, Santa Barbara (UCSB) and Prof. Andrzej Weber, Department of Anthropology, University of Alberta, Canada for their critical remarks that helped in improving the manuscript. I would like to thank Shin Joon-Ho for the statistical advice and for the English corrections suggested by Paedar.

REFERENCES

1. L. M. Andrew, 1968Feng-Shui Theory and Practice Investigated by Spatial Regression ModelingThe Journal of Asian Studies253267
2. F. Carre, M. C. Giard, 2002Feng-Shui Theory and Practice Investigated by Spatial Regression ModelingGeoderma241263
3. W. S. Choi, 2001A study of geomantic auxiliary temples and pagodas in the Yongnam region. Study of Korean History of Thought, 17, 169204
4. T. Chris, 2002The role of Feng-Shui: Haymarket priority project [Sydney, Australia] Landscape Australia 1, 1012
5. L. Diana, 2004California assemblyman proposes resolution that promotes building with Feng-Shui. Architectural Record, 4. 48
6. M. J. Earle, 1992Feng-Shui Theory and Practice Investigated by Spatial Regression Modelinga. Ph.D Thesis, University of Minnesota, 270.
7. ESRI (Environmental System Research Institute), 2006ArcGIS software help menu (spatial analysis toolbox).
8. J. Hammond, 1995Feng-Shui Theory and Practice Investigated by Spatial Regression ModelingModern Asian Studies555572
9. J. S. Hobson, P.1, 1994Feng-Shui Theory and Practice Investigated by Spatial Regression ModelingInternational Journal

of Contemporary Hospitality Management2126

10. W. Hough, 1999Feng-Shui Theory and Practice Investigated by Spatial Regression Modelingologist, 1, 150154
11. X. A. Jun, 2003Framework for site analysis with emphasis on Feng-Shui and contemporary environmental design principles. Ph.D Thesis, Virginia Polytechnic Institute and State University.
12. M. B. June, 2002Feng-Shui Theory and Practice Investigated by Spatial Regression ModelingMaster Thesis, University of South Africa.
13. D. G. Kim, 2002Feng-Shui (Pungsu): chain of life that connects ancestors with descendants. Koreana, 4, 2431
14. Korean Association of Property Appraisers,2006http://member.kapanet.co.kr/cgi-bin/gsv/accessed 10 February 2008).
15. Korean Heritage Online, 2006Buddhist culture in Palgong Mountain, http://www.heritage.go.kr/eng/tou/tou_the_12_sce.jsp.accessed 10 February 2008).
16. S. J. Lu, P. B. Jones, 2000House design by surname in Feng-Shui Source. The Journal of Architecture, 4, 355367
17. E. Lunch, 2003Feng-Shui as a site design tool: Assessing conditions of human comfort in urban places. Master Thesis, University of Arizona.
18. M. Y. Mak, S. T. Ng, 2005The art and science of Feng-Shui- a study on architects' perception. Building and Environment, 3, 427434
19. D. H. Sung, 2001Removing an illustrious imperial garden and the establishment of Hwasong according to Chongjo's view of geography and geomancy, Study of Korean History of Thought [South Korea] 17121167
20. M. L. Tan, 2006Filipino 'feng shui'?
21. http://www.inq7.net/globalnation/col_pik/2003/aug14.htm (accessed 10 February 2008
22. K. Verlyn, 1995Bold illusion for California: unexpected ideas transform a San Jose tract house. Architectural Digest, 5, 156161
23. Xu,P.,1998Feng-shui" models: structured traditional Beijing courtyard houses, Journal of Architectural and Planning Research, 4, 271282

24. H. K. Yoon, 1975Feng-Shui Theory and Practice Investigated by Spatial Regression ModelingAsian Folklore Studies2134
25. H. K. Yoon, 2006The Nature and Culture of Feng-Shui in Korea. Lexington Book
26. N. M. Zetlin, 1995FENG-SHUI: Smart business or supersition? Management Review, 8, 26
27. F. Zitao, 2000Feng-Shui Theory and Practice Investigated by Spatial Regression ModelingMaster Thesis, Arizona State University.

Chapter 4

GIS Applied to the Hydrogeologic Characterization – Examples for Mancha Oriental Aquifer (SE Spain)

David Sanz, Santiago Castaño, and Juan José Gómez-Alday

University of Castilla - La Mancha / Remote Sensing and GIS Group, Albacete,, Spain

INTRODUCTION

The population on planet Earth, according to FAO forecasts, will increase from 6 billion to 8.1 billion inhabitants in 2030 andwill coincide with an increase in water demandsto meethuman needs. Fresh water has ceased to be an inexhaustible resource to become a rather limited and scarce one.

Earth's hydrosphere has an approximate volume of $1.38x10^{10}$ km^3 of water, which has remained virtually constant since its formation over3 billion years ago. This volume of water is distributed into four groups:

- The vast majority is in the oceans, at 97.6% of the total ($1,350x10^6$ km^3),
- In second place is solid water, in glaciers, at 1.9% ($26x10^6$ km^3),
- Third is groundwater, with 0.5% of the total, which is $7x10^6$ km^3, and
- The remainder of water on Earth (0.03% of total) is divided among lakes (0.017), soil (0.01%), the atmosphere (0.001%), Biosphere (0.0005%) and rivers (0.0001%).

Ocean water is salt water and the glaciers are difficult to utilize because they are located far from major populated areas. Therefore, we find that groundwater is the largest volume of freshwater available to man. The volume of groundwater is 4,000 times greater than that of rivers and 30 times higher than the rest of liquid water that is on the surface of the continents. In addition, groundwater has characteristics that make it especially attractive to combating the processes of drought and desertification. Unlike surface water, groundwater does not evaporate, there are no major seasonal variations and flow is very slow, so itis difficult to contaminate (Castaño et al., 2008).

Groundwater is held in a naturally occurring reservoir called an aquifer, a geologic formation capable of storing, receiving and transmitting water so that man can easily take advantage of economically significant quantities to meet needs. The water is contained in any geological formation (ie: river gravel, certified limestone, porous sandstone and so on).

Like all scarce resources, groundwater management must be approached from a dual approach (Knowledge and Sustainability):

Knowledge: There must be a sound understanding of the hydrogeological aquifer system to be managed. This should include a detailed analysis of the hydraulic aspects (geology, hydraulic parameters and groundwater flow) and should be contrasted with the hydro chemical aspects of water containing (origin of the substances dissolved in groundwater and hydro-chemical changes due to movement through groundwater flow).

Sustainability: Groundwater pumping of a region (an aquifer) shouldn't exceed the water received, that is, the available resources. If users pump a volume of groundwater for short-term needs (ie: drought conditions), beyond the resources available, they use the aquifer reserves. Then, the aquifer must be given time to recover (either by saving water or allowing recharge to increase during periods of more rainfall). Otherwise resources suffer overexploitation, putting the aquifer at risk of becoming depleted. It is obvious that the water volumes involved should be determined as precisely as possible.

In managing groundwater resources, Geographical Information System (GIS) are tools capable of storing and managing spatial hydrogeological data by spatial referencing in digital formats. The correlation of all data with location is the key feature of GIS, which provides the ability to analyze and model hydrologic processes and produce results in maps and in digital formats. Thus, GIS can be considered a support system in decision making and an ideal tool for monitoring certain hydrogeological processes with socio-economic impacts (Goodchild et al., 1996).

Figure 1 shows a diagram of a GIS aquifer system modeling tool (Case study of the Mancha Oriental System). This scheme is integrated into a) a block of hydrogeological data maps of the study area, which supplies data on groundwater from urban and industrial and general hydrological information (surface and groundwater hydrology) necessary to carry out the integration and interpretation of some results, and b) a block of data from remote sensing imagery. Remote sensing allows for classification of crops and their relationship to water supply for irrigation, mapping of wells, and the assessment of recharge by precipitation of rainfall. All this information is transferred to software that simulates groundwater flow in the Mancha Oriental aquifer using intersection tools.

The main goal is to show the methods some GIS applications have in hydrogeological studies. This chapter is divided into two sections which describe some examples of hydrogeological characterization, and secondly, a method for calculating groundwater abstraction. To demonstrate these applications one of the largest aquifers in southern Europe (in terms of area), the Mancha Oriental System has been chosen.

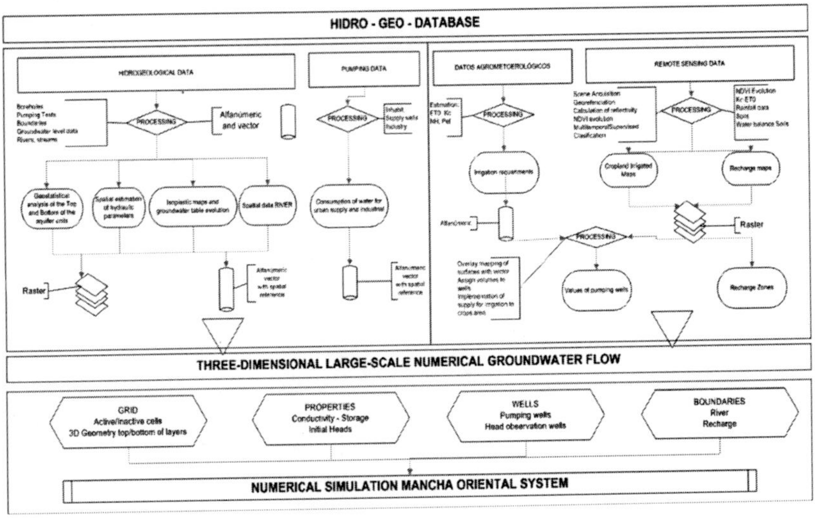

Figure 1: Basic diagram of coupling remote sensing and GIS techniques with the groundwater flow model of the Mancha Oriental System.

GIS & HYDROGEOLOGICAL CHARACTERIZATION

In general, among the Earth Sciences and particularly hydrogeology, sources of data tend to be from points (wells, points of water, lithological columns, etc.) defined by a geographic location (UTM or geographic coordinates) and attributes (topographic top or bottom of a hydrogeologic entity, groundwater level, hydraulic parameters, concentration of a chemical compound). This type of data, usually measured in the field, must be spatially distributed in a continuous manner such that a value is given for any point within the space. To achieve this, interpolation or spatial estimation is used. This method derives an interpolation function that provides estimates for a point in space based on the points measured. GIS tools have incorporated algorithms which perform these operations with discrete entities (vector) and generate spatially continuous entities (raster, line models, etc.). In addition to expanding and the geodatabase and adding values, these techniques create a foundation for spatial modeling (Peña Llopis, 2006).

The most commonly used spatially continuous entities are raster maps, which are characterized by a two-dimensional numerical matrix or digital image. Each element of the matrix, called a picture element or pixel, has an attribute assigned to it in the database. The only requirements are for maps to have attribute values referenced to the same coordinate system and the same number and arrangement of pixels to perform algebra operations with them (ie: isopaches: difference between top and the bottomraster maps of the geologic formation; calculation of storage volumes: difference between raster maps and contour lines or groundwater for different dates, multiplied by the storage coefficient, etc.).

Theoretical Foundations

Using a spatial domain and a series of points (which we will refer to as points of observation) Pi, where i= 1, 2, ...n, which have a series of coordinates xi where variable Z has been measured in Pi points; Zi (observed values), the interpolation or spatial estimation aims to find the value of Z (estimated values) at any point in the known space. An interpolation function must be obtained:

$$Z(x)=\theta(x,xi,Zi) =f(x) \qquad ()$$

The interpolation function should have certain characteristics: a) accuracy: the estimated value in points of measurement should coincide with the measured value, b) spatial continuity, c) ability to be derived: the interpolation should be "smooth" and d) it should be stable with respect to the location of the variable as well as its value such that small variations in data do not provoke large variation in the interpolation. As a function of these characteristics, especially the last condition, there is no universal interpolator and there is always another interpolation method which can be applied (Samper & Carrera, 1996). Most GIS software presents two interpolation methods: Deterministic and Stochastic.

Deterministic Methods

This type of method is characterized by associating a mathematical function, such as an interpolation function, to the measured or observed values, in which these points are considered without error. Following the nomenclature followed until now, this mathematical function could be written in the following manner:

$$Z(x) = f'(x) = \sum_{i=1}^{n} c_i f(x)_i \tag{1}$$

Where for each x a Z(x) value is measured through a function f(x), which is defined by the sumof all "n" points of observation of a product between a base function f(x)i and coefficients, Ci. For example, in a simple exact interpolation the observed or measured values (Zi) coincide with the Cs values, multiplied by a weighting factor given by the function f(x)i. The deterministic interpolation functions differ from one another in the means of evaluating f(x)i and Ci.

There are various deterministic interpolation techniques. The most commonly used methods are presented here (ESRI, 1997) (Samper & Carrera, 1996):

Nearest. neighbor (Thyessen polygons, Polygons of influence).

This method assigns the value of each measured or observed point to each pixel or node of the interpolated area. For each point of observation the Euclidean distance is calculated for all other points and each is given the closest value. The result is a map of polygons with an interpolated value (Fig. 2A). This method is often used for regular grids and/or dense observed data, or to find areas of influence.

Interpolations. based on weighting functions.

The estimation or interpolation in this type of method is performed by a weighted average of the observed values. At each point of observation a weight is assigned. The selection criterion is that the weighting function is exclusively dependent upon the distance (d). The weight will decrease with increasing distance between points. The most common strategy for generating this criterion is the Inverse of Distance raised to some exponent (a).

$$f(x) = \frac{1}{d^a} \tag{2}$$

This exponent shows the "speed" with which the weight of a point of observation decreases with distance from the point of estimation. At times the number of points of influence is restricted, or a radius or maximum distance is assigned for considering points of observation. This interpolation method is exact and is commonly applied, with the only disadvantage being creating the feared "bulls-eyes" (Fig. 2B).

Polynomial Interpolation

In this method the interpolating function is a polynomial function which varies in its exponential order. The choices for polynomial are: a) through exact fit and b) fit by least mean square. The first method aims to resolve the system of equations defined by the n points of observation. If there are many points of observation, the fit of higher order polynomials can become unviable, giving unrealistic interpolations with exaggerated variation among the values (Fig. 2C). In fact, by default these methods limit the polynomial to third order and only use the number of points in a nearby group. One special case of polynomial interpolation is linear interpolation, wherein the interpolation function is a first order polynomial which directly depends on the position of the observed values. It is an exact method and does not take into account the spatial distribution of the variable, with the result of soft surfaces. It is an easy method and is often used, above all in cases when not a lot of data is available and the aim is to study the spatial variation of a certain variable. In general, this interpolation method is not used for spatial estimates on realistic structures (topography, groundwater levels, etc.) but rather to determine the tendency of data (Fig. 2C).

Spline Functions

Within polynomial interpolation, this general method generates a different series of expressions for each subdomain into which the whole interpolation space has been divided, wherein continuity requisites are imposed, especially in the contours common to more than one subdomain. The results of this interpolation tend to be surfaces with small changes in levels (Fig. 2D).

Stochastic Methods

This methodology is based on the premise that the variable to be interpolated is a random function associated with probabilistic distribution laws. This type of method gives a measure of error of the interpolation based on the data. There are two classes of stochastic interpolation: a) non-parametric, which are not exact because the errors are assumed to be independent and b) parametric, wherein the interpolation function depends on certain parameters calculated as a function of the observed data (IDW or Krigging) (Samper & Carrera, 1996). The most common method, Krigging, which is available in most GIS software packages, is explained below.

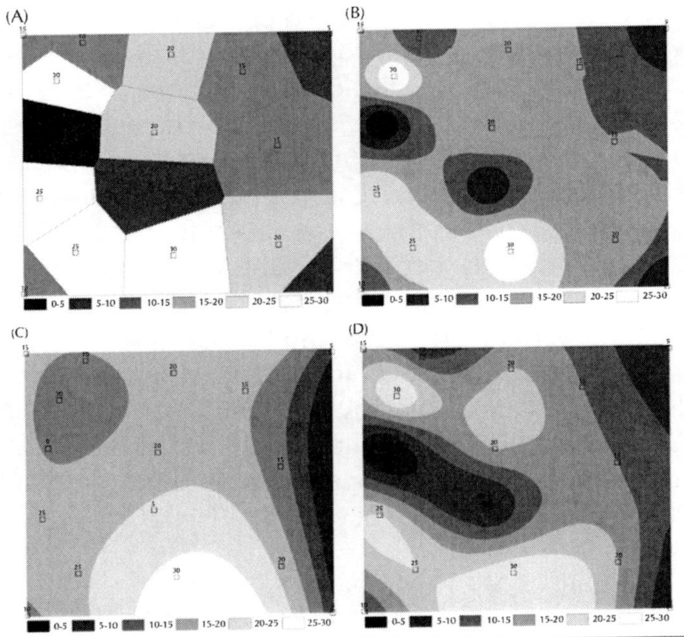

Figure 2: Examples of the different deterministic interpolation methods using GIS tools. A) Polygons of Influence (Thyestean polygons, Nearest neighbor),

B) Interpolations based on weighting functions (in this case the function is the inverse distance squared (IDW); C) polynomial interpolation, in this case third order; D) Spline functions. The same data are used for each case and the results are shown as continuous identical spatial entities,with the same size and pixel size.

Krigging was created under a new discipline, geostatistics, as a resultof problems presented by deterministic interpolation in Earth sciences due to the uncertainty and variability of data (Cassiraga, 1999). The starting hypothesis of geostatistics is that the data of study has a correlation spatial structure, as the realization of an infinite amount of possible realizations. For this reason, geostatistics is called the science of regionalized variables. For spatial estimation using Krigging, the steps below should be followed, among others, and the variable to be interpolated should meet the criteria of normality and stationarity (Johnston et al., 2001).

The first step is structural analysis, with the objective of estimating the semivariogram. This relates the Euclidean distance among the points of observation with the variability of the measured values (Samper & Carrera, 1996). First, the variable should be defined as stationary, if there is a tendency among the data, etc. The function of the semivariogram (estimator of spatial variability) is expressed as:

$$\gamma(x) = \frac{1}{2N}\sum_{i=1}^{n}\left[Z(x_i + h) - Z(x_i)\right]^2 \qquad (3)$$

Where:

Z (xi): experimental data,

h: distance between points of observation (Variogram step),

N: number of pair separated by vector h found in a group of data,

Xi, xi+h: experimental points in an n-dimensional space.

At first, from the observed data an experimental semivariogram will appear. A theoretical function with similar behaviourcan be fit to this in order to calculate a weighting matrix for each point, and statistical error affecting the interpolation can be calculated. The semivariogram is composed of a series of elements (Fig. 3A):

Range: the distance from which the spatial correlation is practically null (Area of influence),

Sill: value that the semivarogram takes in the Range,

Nugget: value of the semivariogram when it intersects at the coordinate axis.

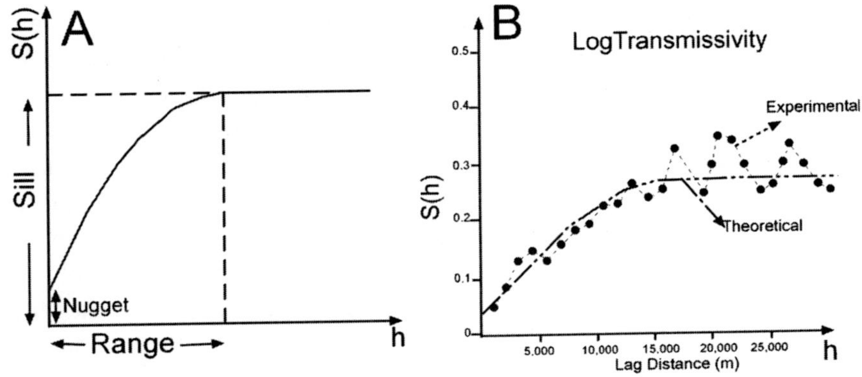

Figure 3: A) Elements of a spherical semivariogram. B) Experimental and theoretical semivariogram sets.

The experimental variogram cannot be used for the geostatistical application. It must be fit to a theoretic model (Fig. 3B). There are different technical variogram models available, with the most popular being the stationary or spherical semivariogram. Once the theoretical semivariogram has been chosen, the Krigging technique performs the spatial estimation of the data. There are diverse Krigging techniques as a function of diverse methodological hypotheses:

Simple: Hypothesis of stationary variable with a known mean and covariance,

Ordinary: Hypothesis of stationary variable with anunknown mean and known covariance,

In an environment (by blocks): Quasistationary hypothesis,

Residual: Non-stationary hypothesis with a known drift, from which residuals are derived and ordinary Krigging can be performed,

Universal: Non-stationary hypothesis and polynomial form with a drift set *a priori*.

MOS Case Study

The Mancha Oriental System (MOS) is located in the SE of the Iberian Peninsula and is one of the largest aquifers in Spain (7,260 km^2) (Fig. 4). The area has a semiarid Mediterranean climate. Average rainfall is 350 mm/year and mean annual temperature is 13-15 C; the continental nature of the climate is clearfrom the extreme temperatures that occur.

The area is characterized as a high plain (700 masl mid-altitude) surrounded by gentle relief, interrupted only by a valley which was carved by the action of the JúcarRiver. From a hydrogeologic perspective, the MOS is formed by the superposition of three limestone aquifer hydrogeologic units (UHs): UH2: Tertiary, UH3: Upper Cretaceous and UH7: Middle Jurassic. These HUs are separated by aquitards/aquifuges that comprise UH1 (upper and lower), UH4, UH5 and UH6 (Sanz et al., 2009). The impermeable base and southwest boundary of the area of study is composed of marl, clay and gypsum from the Lower Jurassic, belonging to HU8 (Fig. 4).

Over the last 30 years the progressive transformation of approximately 100,000 ha from dry to irrigated farmland has translated into an acceleration of socioeconomic development due to widespread use of groundwater resources. Groundwater abstractions in the MOS exceed 400 Mm^3/yr, of which 98% is used for irrigated agriculture and the rest to supply a population of 275,000 Inhabitants (Estrela et al., 2004). Groundwater pumping is not sustainable with the amount of available resources, estimated at 320 Mm^3/yr by the Júcar Water Authority. Therefore, two major impacts are occurring: (a) the quantity of available groundwater is descending, noted as a continuous decline in the regional water level and a decrease in aquifer discharge to the JúcarRiver; and (b) the quality is also affected, as researchers have found a significant increase in nitrate concentrations in groundwater (Moratalla et al., 2009).

In this context, the MOS is an ideal case study for testing and validating the usefulness of GIS Techniques for understanding the aquifer system and planning for sustainable management.Following is a description of the interpolation methods applied to these variables: a) The elevations of the top and bottom of the aquifer units, b) Hydraulic parameters, c) Groundwater level data d) Groundwater chemistry. The approach is to explore the variable data with histograms and spatial trend analysis in orderto understand the behaviourof the variable in space as well as to establishwhether the data are consistent or anomalous. After analyzing the data, a variable is selected to be interpolated. The type of interpolation to apply for raster maps or continuous spatial entities is chosen.

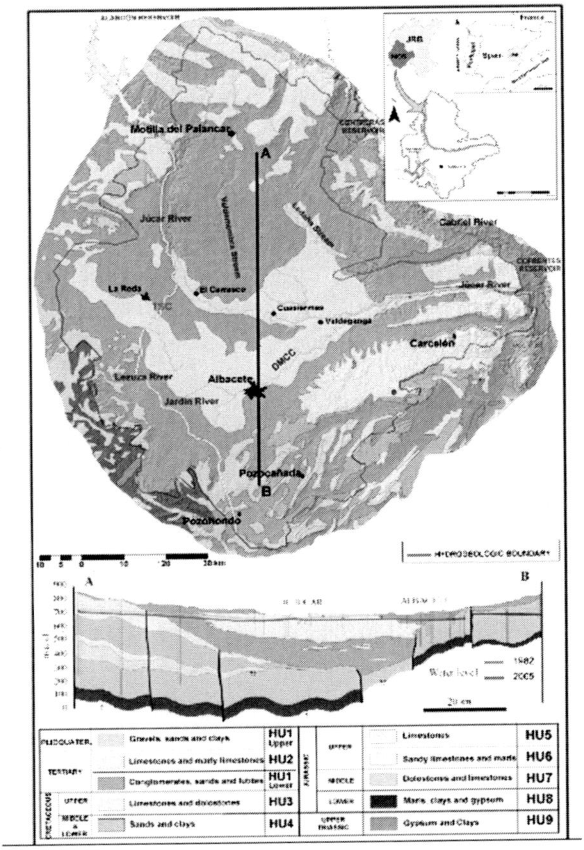

Figure 4: Simplified hydrogeological map of the MOS. Taken from Sanz et al. (2009). JRBJucarRiver Basin.Graphical output from GIS software.

Hydrostratigraphic Framework

Any attempt at making a coherenthydrogeological model should be approached by first understanding, with a certain amount of precision, the geometric configuration. In addition to information on the surface geology, lithologic columns should be analyzed and gathered from water sampling points, and tested materials should be classified within the defined hydrogeologic units (Murray & Hudson, 2002). Using this information, layers of geographically located points (X, Y coordinates) as well as the topographic height of the superior (top) and inferior (bottom) limits were made into attributes of each hydrogeologic unit that behaves as an aquifer. Using geostatistical interpolation models developed on theoretic and applied foundations, GIS software (i.e. ArcMap ® 9.3) was used to determine the continuous geographic entities, ie: raster maps of the surfaces corresponding to the top and bottom of each hydrogeologic aquifer unit. The result is the three-dimensional structure of the hydrogeologic system (Fig. 5A). These 3D geologic models (Fig. 5), constructed using GIS tools, became the foundation for the numeric simulation models in later steps.

Hydrodynamic Characterization

Transmissivity, permeability and storage coefficient are hydraulic parameters that must be quantified for an aquifer because they are neededto estimate the progression of groundwater levels, groundwater flow through a section of the aquifer, contaminant transport time, the degree of aquifer homogeneity and the numeric parameterization of the groundwater flow models (Mace, 2000). Generally, estimating these parameters requires pumping tests in specific points. These provide specific geographic entities defined by their coordinates, and attribute values for the hydraulic parameters in the well. It is also useful to have previous knowledge on the spatial behavior of the variable and establish a relationship for the interpolation model (ie: the type of distribution function of the variable). In the case of the Mancha Oriental System, to determine the spatial distribution of any of the parameters mentioned, these logarithms have been used because the variable tends to have a log-normal distribution. In this case, the value estimated by the Krigging method is the absolute optimum and the semivariogram better represents the structure of the spatial variability (Samper & Carrera, 1996). Once the structure of the spatial variability of log-T was studied, ordinary Krigging type interpolation models were applied (Fig. 5B and 3A).

Characterization of Ground Water Flow

As is the case with aquifer hydraulic parameters, data on the height of the groundwater levels are also point data. The attribute of the groundwater level in this point geographic entity is compiled in the inventory of water points by subtracting the topographic height of the point from the depth of the water level in the well. These measurements should be performed for a specific date and in the least amount of time possible.

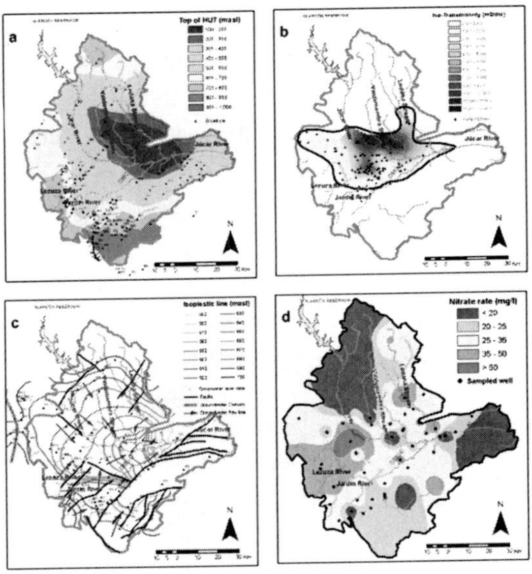

Figure 5: Examples of hydro geologic variable interpolation. a) The elevations on the top of HU7Mid-Jurassic aquifer, b) Hydraulic parameters (Transmissivity of HU2 Miocene Aquifer,) c) Groundwater level data for 1975 d) Groundwater chemistry (Nitrate rate-2008).

The raster maps obtained using the data on groundwater level height are called groundwater contour maps (isopiestic lines). These maps serve to determine how groundwater flow functions, where are the recharge and/or pumping areas in addition to indicating gradient calculations, flow and permeability (Fig. 5C). By crossing groundwater contour maps for different dates, contour descent maps can be obtained for that period. In addition, Variation in Saturated Thickness (VST) can be calculated between those dates.

Hydro chemical Characterization

The chemical composition of groundwater is conditioned by a multitude of factors. Among those, the most important are: a) chemical composition and disposition of materials with which the water is in contact, b) time of contact with these materials, c) temperature, d) pressure, e) presence of gases, and f) level of water saturation in relation to distinct incorporated salts (Custodio & Llamas, 1983). Although the composition of groundwater is continually changing, the anthropogenic factors can significantly influence the composition. In fact, changes in land use are considered the most influential factor in groundwater pollution. Ions such as NO_3, SO_4, Na and Cl can come from agricultural fertilizers, livestock waste and waste from industry and urban centers. Nitrate is accepted as the most common contaminant in groundwater (Gulis et al., 2002; Jalali, 2009).

In Europe, the objective is for waterways to achieve "good" chemical and ecological status according to Directive 2006/118/EC of the European Parliament and Commission (DOCE, 2006). This directivedescribes the protection of groundwater from pollution and deterioration and the establishment of a pollution prevention and reduction plan by 2015. In addition, water bodies should be in good quantitative and qualitative status, especially in reference to nitrate content, which should not exceed 50 mg/l.

Thus, establishing the spatial distribution of NO_3 concentrations in groundwater within the aquifer is of vital importance. To accomplish this, point analyses of groundwater in wells and springs must be performed. Using advanced interpolation capacities provided by GIS tools, a complete geostatistical study can be performed to establish the most contaminated areas in terms of nitrate (Fig. 5D).

GIS & GROUNDWATER ABSTRACTIONS

Intensive use of groundwater for irrigation in arid and semiarid regions has often been the main driver of socioeconomic development over the past four decades (Shah, 2005). However, poor management of pumped volumes of water has led to negative consequences in terms of quality and quantity of available groundwater resources and associated ecosystems.

Controlling the groundwater withdrawals from a wide area of intensive irrigation is not easy. The largest volume of water used for agriculture has been extracted through tens of thousands of pumping-wells which generally have no measurement system and, in many cases, do not meet legal requirements or are unknown even in their location. Various methods of calculating groundwater abstractions have been known for years, but all of themare very expensive or inaccurate in their application to large areas (Brown et al., 2009).

In this scenario, the data provided by satellites (remote sensing) and the computerized processing of these geo-referenced data (GIS) represent a new approach to monitoring and quantifying groundwater abstractions, with the following characteristics: instantaneous observations are available over large areas, there are several images throughout the year, there is information not visible to the naked eye, data distributed in both space and time is available, the information is not conditioned by the legal or administrative characteristics of the pumping wells, and satellite image acquisition and processing is very low-cost compared to traditional methods (Castaño et al., 2009).

Theoretical Foundations

The methodology for determining groundwater pumping for irrigation follows these steps: First, the irrigated crops are identified and classified by the multitemporal analysis of images obtained by multispectral sensors on satellite platforms, comparing the phenological evolution of the crops with the evolution of the Normalized Difference Vegetation Index (NDVI; González-Piqueras, 2006). Then, the area covered by crops is quantified by introducing the data into a Geographic Information System (GIS) and overlay them with the areas or required limits. Based on the surface area of each crop and the knowledge of their water requirements, the theoretical amount of water needed for those crops to reach the stage of development seen in the images is calculated. When agricultural practices are known, a correction factor is applied to translate the theoretical amount of water applied to each crop. Finally, all the information generated is integrated (spatially and temporally distributed) in a Geographic Information System (see Figure 1) and is used to establish relationships among all elements of the water balance (Brown et al., 2009).

Multitemporal Analysis of Satellite Images and Cross with Vector Cartography

The term "Remote Sensing" has different definitions, but the most commonly used is "a group of techniques that analyze data obtained by multispectral sensors located on airplanes, spatial platforms or satellites." The sensors (on satellites) that observe the surface of the Earth are instruments that register the radiation from Earth and the atmosphere and transform it into a signal that can be managed in analog or digital format (Calera et al., 2006). The sensors do this by detecting the electromagnetic signal from the Earth and the atmosphere of a certain wavelength and converting them into an established physical magnitude. The energy values detected, quantified and coded from the sensors are usually in a two-dimensional number matrix or digital image (raster). Each element of the matrix, called a picture element or a pixel, has a digital value assigned to it (digital levels) which is usually registered in a byte or binary code (2^8 values, from 0 to 255). These represent the energy associated with the wavelength to which the detector is sensitive.

According to Chuvievo (2002), each satellite scene can be used to extract four types of information, each with its respective resolution (Table 1):

- Spatial, derived from the organization and presence of elements on the surface of Earth in three dimensions,
- Spectral, dependent upon the observed and measured energy,
- Temporal, associated with changes over time in a specific spatial location and
- Radiometry, related with the conversion of voltage collected by the apparatus that receives the signal sent from quantifying entities and later on digital levels.

Table 1: Type of resolution of two kinds of sensors, such as Landsat and NOAA satellite (see http//:www.landsat.org and www.noaa.gov/)

Landsat Satellite TM sensor				**NOAA Satellite AVHRR senso**r			
Spatial resolution, pixel size	Spectral resolution	Temporal resolution	Radiometric resolution	Spatial resolution, pixel size	Spectral resolution	Temporal resolution	Radiometric resolution
30 x 30 meters	7 bands	15 days	8 bits	1x1 Km	5 bands	12 hours	10 bits

The information referred from the sensor is treated digitally to obtain a geo-referenced representation of the land. Once the interactions of the atmosphere are removed from it, the radiation values received correspond exactly with those measured on the surface (see more detailed information in Chuvieco (2006) or Calera et al., (2006).

The source of radiant energy is solar radiation on the land surface after moving through the atmosphere. The radiation the sensor obtains is that which emerged from the land surface to the proper region of the spectrum when the emissions due to temperature are considered null. Therefore, the electromagnetic spectrum is the continuous succession of these frequency values (wavelengths). Conceptually it can be divided into bands in which electromagnetic radiation has a similar behavior (Fig. 6).

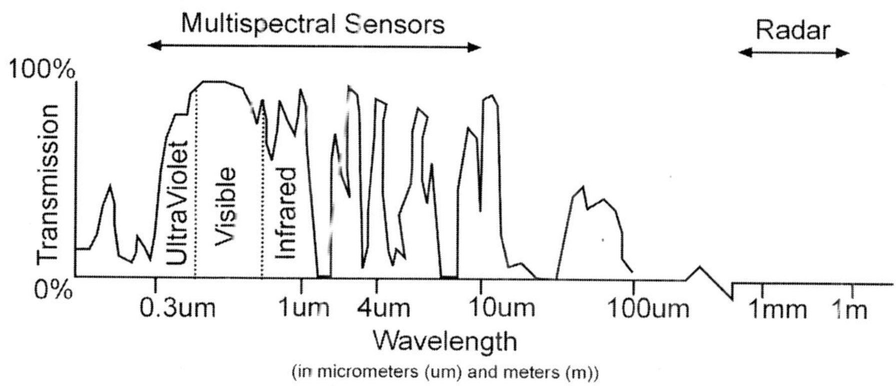

Figure 6: Main features of the electromagnetic spectrum.

Three basic elements can be distinguished as the components comprising all forms of the landscape on the Earth's surface: soil, water and vegetation. The behaviour of these elements in different regions of the electromagnetic spectrum can be observed in Figure 7. The energy emitted (reflectivity) from the ground in the solar spectrum has a uniform response, showing a flat curve and ascending to greater wavelengths. It is important to know that bare soil can present different curves according to the chemical composition, humidity content, organic material content, etc. In the optical spectrum, water can be observed as a strong contrast between the reflectivity of the visible (5%) and the infrared, where water absorbs almost all the radiation in these wavelengths (Fig. 7). This effect is used to separate the water-soil limit. Similarly to soil, the characteristic reflectivity curve for water can vary as a function of factors such as depth, suspended materials, roughness, etc. (Calera et al., 2006).

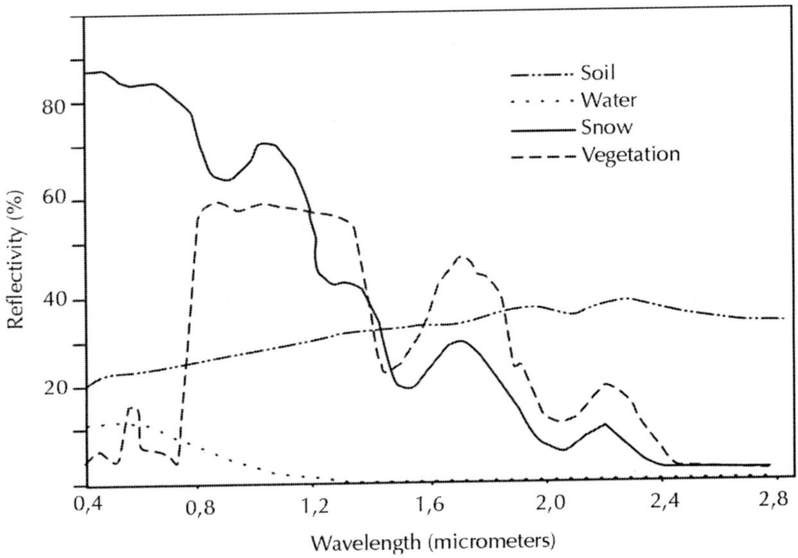

Figure 7: Reflectivity curve characteristic of soil, water and vegetation

The morphology of the reflectivity curve against the wavelength that vegetation has is well defined (Fig. 7). It has low reflectivity (10%) with a maximum relative to the region of green, high reflectivity in the near infrared which is gradually reduced to the middle of the infrared spectrum. The strong contrast between the reflectivity of red and near infrared indicates that the higher the contrast is, the more vigorous the vegetation is, either due to greater land cover or greater photosynthetic activity (Calera et al., 2006). This spectral behaviouris the foundation for the development of certain indices with an objective to highlighting active vegetation from other components (soil, water, dry vegetation) From the reflectivity of each band (quantitative information distributed and geo-referenced in space) a relationship with the biophysical characteristics can be established (biomass, fraction of plant cover, etc.). This allows for quantitative, spatial-temporal monitoring of the processes on the Earth's surface (Bastiaanssen et al., 2000; Calera et al., 2001; González-Piqueras, 2006). Nonetheless, reflecting the spatial and temporal variability of plant cover is complicated if different spectral bands with the reflectivity values are used. To unify this process, the Vegetation Indices have been developed, one of the most important being the NDVI (Rouse et al., 1973).

The NDVI is:

$$NDVI = \frac{(NIR - R)}{(NIR + R)} \tag{4}$$

Where:

NDVI: Normalized Difference Vegetation Index,

NIR: Near Infrared reflectivity (spectrum range in micrometers),

R: reflectivity in red reflectivity.

GIS tools have the capacity to establish dynamic processes if they contain spatially referenced information which is repeated over time in addition to the ability to study spatial changes over the land surface.Mathematical operations can be used between the different sensor bands(digital images) to obtain quantitative information of each satellite scene obtained for a specific date.In this way, a temporal series is available for establishing the progression of a variable, for example NDVI.Multitemporal analysis stems from the availability of a time sequence of images, so these scenes must meet a set of requirements such as geometric coregistration (ability to superimpose images with the highest precision possible and radiometric normalization; Calera et al., 2005).

With this information and digital classification tools each pixel of the image from each date can be assigned a class defined through an automated process.There are two methods for classification: a) supervised and b) unsupervised (does not require intervention of an "interpreter").The difference between the two is in the method of obtaining the spectral reference classes for assigning one to each pixel.

Supervised classification stems from *a priori* knowledge on specific land uses located in space, which are called training plots. These serve to establish spectral reference classes. There are several methods and a procedure for assigning a class to each pixel, but the most commonly used is an algorithm of maximum probability. Without getting into the details of this method, the algorithm is based on multivariate statistical analysis of components that identify each pixel in terms of their closest resemblance.

Other classification methods that could be used as alternatives or complimentary methods are decision tree (expert systems). These procedures are based on separating the pixel values of a layer into homogeneous groups and subgroups. Another method, called contextual filters, can also be applied. This not only considers the spectral characteristics of an individual pixel, but also considers neighboring pixels (Calera et al., 2006).

Once the classified map has been obtained, the spectral classes can be used to select the classes that are of interest from a hydrogeologic point of view.In our case study, this is crops irrigated with groundwater. Therefore, it is important to know the area of irrigated crops and their spatial distribution.One of the most commonly used GIS techniques for this is overlay vectorial and raster cartography, which is the only way to obtain this information for rasterized areas.There are two types of overlays that dependon the pixel value (ESRI, 1997). When the pixel has a real value (for example a precipitation map, groundwater level, NDVI, etc.) a statistic is calculated for areas by obtaining statistical values from the raster within the selected polygons (mean, minimum, maximum, etc.). The other case is when the attribute of each pixel is a discrete value defining a series of classes (i.e. raster map of the classified land uses). In this occasion (tabulate areas) the result is the surface of each class within the vectorial polygons selected (ESRI, 1997).Once the areas of each irrigated crop have been determined and the amount of irrigation water supply is known, it is possible to calculate the volume of water used to irrigate crops in the area on an annual basis (see summary in Castaño et al., (2009).

MOS Case Study

The use of groundwater resources of the MOS above its recharge capacity has led to several quantitative impacts: a steady decline in groundwater level, reduced aquifer discharge to the Júcar River and aquifer pollution. In fact, the quantitative analysis performed on the Júcar River Basin (Estrela et al., 2004) for the European Water Framework Directive (EU, 2000) clearly indicates that the environmental objectives set are not being reached at the present time and there is a certain risk of not meeting them by 2015.

In this situation, quite common in a semi-arid river basin, it is particularly important to precisely quantify the groundwater balance in order to determine aquifer sustainability. The information provided by the multispectral images becomes critical because these data sets are the only consistent and objective information on crops and can replace the data on agricultural statistics. In this regard, the MOS is an ideal case study for testing and validating the adequacy of remote sensing and GIS techniques for calculating groundwater abstractions in agricultural basins in semi-arid climates (Castaño et al., 2009).

Following is a description of several studies in the MOS to classify irrigated crops and quantify the ground water consumption required for ideal phenological development.

Calculation of Groundwater Withdrawals

The development of a method to calculate groundwater abstractions has been briefly described in the section on the theoretical foundations. In addition to knowing the method, one must have previous knowledge of the study area in order to choose the type of satellite image, for example crops and natural vegetation (phenologic development). soils, climate, relief, etc. In this study, due to the characteristics of the study area, the ideal images for thematic cartography are those from Landsat5 TM and Landsat7 ETM+ (Tables 1 and 2).

Table 2: Bands and wavelengths from the Landsat 5 TM and Landsat 7 ETM+. Information on band 6 from this sensor is omitted because it is not relevant in this study

	Band 1	**Band** 2	**Band** 3	**Band** 4	**Band** 5	**Band** 7
Wavelength (λμm)	0.45	0.52	0.45	0.52	0.45	0.52
Region	Blue	Green	Red	Near Infrared	Middle Infrared	Far Infrared

Using this information, the number of scenes required can be established as well as the bands to use for differentiating the crops of interest. For example, at least two images are necessary to establish a time series and to identify the non-irrigated crops, one in May or

June (maturation process) and another in July (harvest). If more Landsat scenes on specific dates are included spring irrigated crops can be identified. Generally, a minimum of 16 images are used for performing the classification. If the temporal progression of the spectral response is considered a discriminating element for crops (phenologic development), the NDVI spectral band is the most useful. This is obtained by performing mathematical operations with the images for the same dates using bands 3 and 4 of the Landsat sensor (Table 2 and Figure 8).

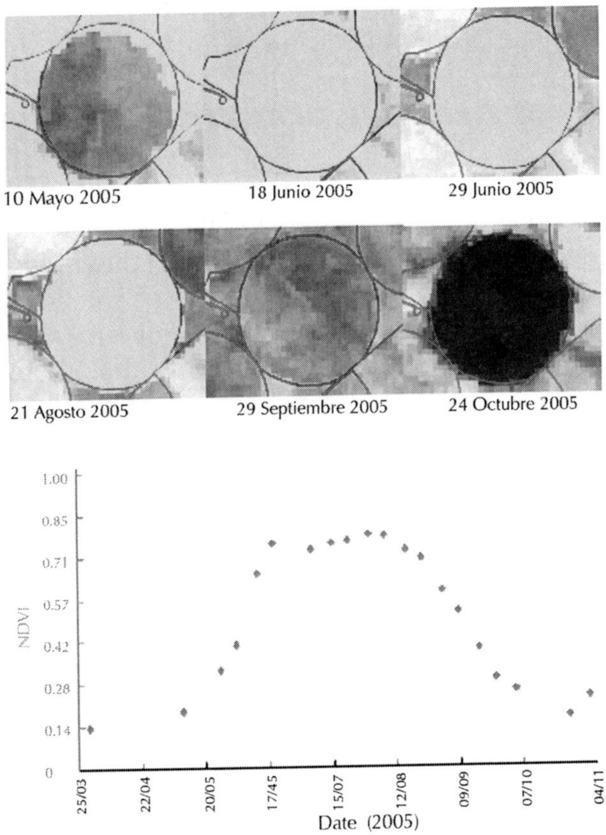

Figure 8: Example of the RGB composition for a maize plot and the temporal progression of NDVI during the crop growth cycle in the year 2005. (From Calera, et al., 2005).

The next step is to choose the classes to use in the classification as a function of those which can be differentiated using the spectral band in the images used such that they meet the objectives of the study (Spring Irrigation, Summer Irrigation, Spring-Summer Irrigation, Alfalfa, Bare soil, Dry farming crops, Shrubs, Forest).For classification, a study is required on the training plots called "true land."

This information is used to perform the classification using the maximum probability algorithm, tree decision and contextualfilters. These classifications will be capable of discriminating between the sources of error that can be generated, e.g. in dry farmed crops because several images can have similar spectral responses. For example, in a rainy spring, cereals grown in irrigated or dry farming can be difficult to differentiate. Therefore, using other scenes with tree decision can help in the classification process. Contextual filters can be used to eliminate error on the plot boundaries or isolated pixels that belong to a different class than the rest of the plot. The result is a raster map with the classification of all irrigated crops (Fig. 9).

Figure 9: Cropland irrigated with groundwater map, classified by remote sensing techniques for a2001 in the MOS from LANDSAT-TM scenes.

Once the maps have been classified, the spectral classes can be used to select the classes of interest from a hydrogeologic point of view. The area of irrigated crops must be determined (divided into spring, summer and spring-summer irrigated crops) as well as their spatial distribution (overlay tools). Information on the irrigation needs of the crops present in the MOS are provided weekly through the Irrigation Assessment Service (SAR) by the Agronomic Institute of Technology of the Province of Albacete (ITAP) using the method proposed by Allen et al., (1998). For each agricultural year, the institution groups the irrigation needs for each crop and publishes them in the annual monitoring reports (http://www.itap.es). The theoretical irrigation needs represent the minimum water consumption for sustaining the crops of interest. To determine the true water needs, correction coefficients must be applied to the theoretical irrigation volumes (Castaño et al., 2009). Field work to quantify the agricultural practices applied in the region should be done to perform this calculation. Thus, the irrigation efficiency can be used to calculate the correction coefficient that transforms the theoretic quantity of water necessary into the true values applied to each crop in the area. Generally the true amount of groundwater abstraction is higher than the theoretical irrigation needs.

Therefore, the calculation of water consumption for the different types of irrigated crops by applying the following equation:

$$Vr = A_i \times D \tag{5}$$

Where:

Vr = the annual volume of irrigation water for each type of crop (m^3).

A = the area of each type of irrigated crop (ha).

D = the irrigation needs for each type of crop, applying the correction coefficient (m^3/ha).

i = Hydrogeologic Domain, Municipality…In this way groundwater consumption could be calculated for the MOS or any part of it.

Estimating the amount of water required for irrigation is critically important in times of water shortage and especially in the current situation of increasing water demands with increasing populations. The use of GIS tools in this endeavor greatly increases the accuracy and efficiency of these types of study. This chapter is meant to be a summary of some methods used in the case study of the Mancha

Oriental System, but they can be applied to other systems worldwide at risk of groundwater overexploitation or as a preventative measure to protect natural resources.

ACKNOWLEDGEMENTS

This study was funded by the Spanish Government under research grant CGL2008-06394-C02-02/BTE. Special thanks go to the Jucar Water Authority (CHJ) and Stakeholders (JCRMO) in the Mancha Oriental System for providing information. The findings presented therein belong to the authors.Special thanks go to Dr. A. Moratalla (UCLM) for providing some of the chemistry data and toDr. A. Calera and Dr. M. Belmonte (UCLM), for providing the multitemporalclassification of the crops in the study area. We would also like to thank S.A.Kroll for improvingthe English text.

REFERENCES

1. R. G. Allen, L. S. Pereira, D. Raes, Smith, 1998 GIS Applied to the Hydrogeologic Characterization – Examples for Mancha Oriental Aquifer (SE Spain) FAO Irrigation and Drainage. Paper 56. FAO, Rome 300pp.
2. . W. G. M. Bastiaanssen, D. Molden, I. W. Makin, 2000 GIS Applied to the Hydrogeologic Characterization – Examples for Mancha Oriental Aquifer (SE Spain). Agric Water Manage 46 137155 .
3. A. Calera, A. M. Jochum, A. Cuesta, A. Montoro, P. López, 2005 GIS Applied to the Hydrogeologic Characterization – Examples for Mancha Oriental Aquifer (SE Spain). Irrig Drain. 19. 337353 .
4. A. Calera, C. Martínez, J. Meliá, 2001 GIS Applied to the Hydrogeologic Characterization – Examples for Mancha Oriental Aquifer (SE Spain) barley. Int J Remote Sens. 22. 33573362 .
5. A. Calera, S. y. Castaño, A. Quintanilla, 2006 Historia y visión general de la Teledetección. In La evaluación del impacto ambiental de Proyectos y Actividades agroforestales. La Información en un SIG. In La evaluación del impacto ambiental

de Proyectos y Actividades agroforestales. Ed. Ediciones de la Universidad de Castilla-La Mancha. Cuenca. 8-48427-416-0

6. E. Cassiraga, 1999 Incorporación de información blanda para la cuantificación de la incertidumbre: aplicación a la hidrogeología. Tesis doctoral, Universidad Politécnica de Valencia.
7. S. Castaño, J. J. Gómez-Alday, D. Sanz, 2008 Aguas continentales: Gestión de recursos hídricos y calidad del agua. Teledetección y SIG en la gestión de aguas subterráneas. Centro Superior de Investigaciones Científicas. (CSIC). Madrid. 978-8-40008-664-0
8. S. Castaño, D. Sanz, J. J. Gómez-Alday, 2009 GIS Applied to the Hydrogeologic Characterization – Examples for Mancha Oriental Aquifer (SE Spain). Water Resou Manage 24 795814 .
9. E. Chuvieco, 2006 Teledetección ambiental. 2 Ed. Ariel Ciencia, Barcelona. 592p. 978-8-43448-072-8
10. European Committee 2000 The Water Framework GIS Applied to the Hydrogeologic Characterization – Examples for Mancha Oriental Aquifer (SE Spain)n, Brussels.
11. E.. Custodio, M. R. Llamas, 1983 Hidrología subterránea. Omega, Barcelona. 2350 pp. 8-42820-446-2
12. Environment Systems Research Institute (ESRI) 1997 Understanding GIS- the ARC/INFO Method, Self-study Workbook. 4th Ed.
13. T. Estrela, (ed, 2004 Jucar Pilot River Basin. Provisional Article 5 Report pursuant to the Water Framework Directive. Ministerio de Medio Ambiente, Valencia, Spain.
14. M. F. Goodchild, (ed, 1996 GIS Applied to the Hydrogeologic Characterization – Examples for Mancha Oriental Aquifer (SE Spain) GIS World Books, Fort Collins, CO.
15. G. Gulis, M. Czompolyova, J. R. Cerhan, 2002 GIS Applied to the Hydrogeologic Characterization – Examples for Mancha Oriental Aquifer (SE Spain). Environmental Research; 88 3 182187 .
16. M. Jalali, 2009 GIS Applied to the Hydrogeologic Characterization – Examples for Mancha Oriental Aquifer (SE Spain). Environmental Geology; 56 14791488 .
17. K. Johnston, Hoef. J. M. Ver, K. Krivoruchko, N. Lucas, 2001 Using Arc-Gis Geostatylical Analyst. Ed. ESRI. USA, 300 p.
18. R. E. Mace, 2000 GIS Applied to the Hydrogeologic Characterization – Examples for Mancha Oriental Aquifer (SE

Spain), The University of Texas at Austin, Bureau of Economic Geology, Geological Circular 012

19. A. Moratalla, J. J. Gómez-Alday, Heras. J. De las, D. Sanz, S. Castaño, 2008 GIS Applied to the Hydrogeologic Characterization – Examples for Mancha Oriental Aquifer (SE Spain). Water Resour Manag 29 16211640 .
20. K. Murray, E., M. R. Hudson, 2002 Three-Dimensional Geologic Framework Modeling for a Karst Region in the Buffalo National River, Arkansas. U.S. Geological Survey Karst Interest Group. Proceedings, Shepherdstown, West Virginia. Eve L. Kuniansky, editor. Water-Resources Investigations Report 024174
21. J. González-Piqueras, 2006 Evapotranspiración de la cubierta vegetal mediante la determ,inación del coeficiente de cultivo por teledetección. Extensión a escala regional: Acuífero Mancha Oriental. Dissertation, University of Valencia.
22. Llopis. J. Peña, 2006 Sistemas de Información Geográfica aplicados a la gestión del territorio. Ed. Club Universitario. Alicante, 310 p. 8-48454-493-1
23. F. Samper, J., J. Carrera, 1996 Geoestadística: aplicaciones a la hidrogeología subterránea. 2ª Ed. Mundi-Prensa. Barcelona, 484 p. 8-44046-045-7
24. D. Sanz, J. J. Gómez-Alday, S. Castaño, A. Moratalla, Heras. J. De las, Alfaro. P. M. Martínez, 2009 Hydrostratigraphic framework and hydrogeological behaviour of the Mancha Oriental System (SE Spain). Hydrogeol J 17 13751391
25. T. Shah, 2005 Groundwater and Human Development: Challenges and Opportunities in Livelihood and Environment, Water Science and Technology 8 2737

Chapter 5

The Geospace Response to Variable Inputs from the Lower Atmosphere: A Review of the Progress Made By Task Group 4 of CAWSES-II

Jens Oberheide1, Kazuo Shiokawa[2], Subramanian Gurubaran[3], William E Ward[4], Hitoshi Fujiwara[5], Michael J Kosch[6, 7, 8], Jonathan J Makela[9], and Hisao Takahashi[10]

[1]Department of Physics and Astronomy, Clemson University, Clemson 29634-0978, SC, USA

[2]Solar-Terrestrial Environment Laboratory, Nagoya University, Nagoya 464-8601, Japan

[3]Indian Institute of Geomagnetism, Navi Mumbai 410 218, India

[4]Department of Physics, University of New Brunswick, Fredericton E3B 5A3, NB, Canada

[5]Faculty of Science and Technology, Seikei University, Tokyo, Japan

[6]South African National Space Agency, Hermanus 7200, South Africa

[7]Lancaster University, Lancaster LA1 4YB, UK

[8]Currently at National Institute of Polar Research, Tachikawa, Japan

[9]University of Illinois at Urbana-Champaign, Urbana, IL, USA

[10]Instituto Nacional de Pesquisas espaciais, São José dos Campos, SP, Brazil

ABSTRACT

The advent of new satellite missions, ground-based instrumentation networks, and the development of whole atmosphere models over the past decade resulted in a paradigm shift in understanding the variability of geospace, that is, the region of the atmosphere between the stratosphere and several thousand kilometers above ground where atmosphere-ionosphere-magnetosphere interactions occur. It has now been realized that conditions in geospace are linked strongly to terrestrial weather and climate below, contradicting previous textbook knowledge that the space weather of Earth's near space environment is driven by energy injections at high latitudes connected with magnetosphere-ionosphere coupling and solar radiation variation at extreme ultraviolet wavelengths alone. The primary mechanism through which energy and momentum are transferred from the lower atmosphere is through the generation, propagation, and dissipation of atmospheric waves over a wide range of spatial and temporal scales including electrodynamic coupling through dynamo processes and plasma bubble seeding. The main task of Task Group 4 of SCOSTEP's CAWSES-II program, 2009 to 2013, was to study the geospace response to waves generated by meteorological events, their interaction with the mean flow, and their impact on the ionosphere and their relation to competing thermospheric disturbances generated by energy inputs from above, such as auroral processes at high latitudes. This paper reviews the progress made during the CAWSES-II time period, emphasizing the role of gravity waves, planetary waves and tides, and their ionospheric impacts. Specific campaign contributions from Task Group 4 are highlighted, and future research directions are discussed.

INTRODUCTION

The space weather of Earth's upper neutral and ionized atmosphere is strongly influenced by energy injections at high latitudes connected with magnetosphere-ionosphere coupling and solar radiation variability at extreme ultraviolet wavelengths. A variety of new evidence obtained over the past few years now demonstrates unequivocally that the geospace environment also owes a substantial amount of its variability to waves forced in the lower parts of Earth's atmosphere, that is, the troposphere and stratosphere. This paradigm shift in understanding the causes of geospace variability was made possible by the advent of satellite missions like Imager for Magnetopause-to-Aurora Global Exploration (IMAGE) (Burch [2000]), Thermosphere Ionosphere Mesosphere Energetics and Dynamics (TIMED) (Yee et al. [1999]), CHAllenging Minisatellite Payload (CHAMP) (Reigber et al. [2006]), Gravity Recovery and Climate Experiment (GRACE) (Tapley et al. [2004]), and Constellation Observing System for Meteorology (COSMIC) (Anthes et al. [2008]), advanced ground-based observing capabilities like Poker Flat Advanced Modular Incoherent Scatter Radar (PFISR) (e.g., Nicolls and Heinselman [2007]) and airglow observing networks (e.g., Shiokawa et al. [2009]; Makela et al. [2009]), and the development of 'whole atmosphere' general circulation models from the ground to the ionosphere such as Whole Atmosphere Community Climate Model - thermosphere extension (WACCM-X) (Liu et al. [2010a]), Whole Atmosphere Model (WAM)/Coupled Thermosphere Ionosphere Plasmasphere Electrodynamics Model (CTIPe) (Fuller-Rowell et al.[2010]), and ground-to-topside model of Atmosphere and Ionosphere for Aeronomy (GAIA) (Jin et al. [2011]). See the list of abbreviations at the end of the paper. It is by now without dispute that geospace owes much of its longitudinal, seasonal-latitudinal, and day-to-day variability to meteorological weather processes in the troposphere and stratosphere (e.g., Immel et al. [2006]; Forbes et al. [2006]; Oberheide et al. [2006b]; Hagan et al. [2007]; Goncharenko et al. [2010]; Fritts and Lund [2011]; Maute et al. [2012]; and references therein). The primary mechanism through which energy and momentum are transferred from the lower atmosphere is through the propagation and dissipation of atmospheric waves, including electrodynamic coupling through dynamo processes and instability seeding. Table 1 summarizes the basic characteristics of the three most important classes of atmospheric waves: gravity waves (GWs), atmospheric tides, and planetary waves (PWs).

Table 1: Basic characteristics of atmospheric waves

	Gravity (buoyancy) waves	**Atmospheric tides**	**Planetary waves**
Spatial scale	Local; 10 to 1,000 s of kilometers	Global; 10,000 s of kilometers	Global; 10,000 s of kilometers
Temporal scale	Minutes to hours	1 day or harmonics	Days to weeks
Excitation	Mechanical distortions, latent heating	Solar heating, latent heating	Mechanical distortions, baroclinic instabilities

Oberheide et al.

Oberheide et al. Progress in Earth and Planetary Science 2015 2:2 doi:10.1186/s40645-014-0031-4

CAWSES-II Task Group 4 (TG4; What is the geospace response to variable inputs from the lower atmosphere?) was therefore charged to elucidate the dynamical coupling from the low and middle atmosphere to the geospace (i.e., the upper atmosphere, ionosphere, and magnetosphere), for various wave frequencies and scales, and for equatorial, middle, and high latitudes. As meeting the challenge clearly requires a systems approach involving experimentalists, data analysts, and modelers from different communities, an essential part of TG4 was to encourage interactions between atmospheric scientists and plasma scientists on all occasions. To distinguish geospace variability due to solar and magnetospheric driving from above from processes propagating from below, and in order to dissect the problem into solvable pieces, four projects were formed to respectively address each of the following science questions: (1) How do atmospheric waves connect tropospheric weather with ionosphere/thermosphere variability? (2) What is the relation between atmospheric waves and ionospheric instabilities? (3) How do the different types of waves interact as they propagate through the stratosphere to the ionosphere? (4) How do thermospheric disturbances generated by auroral processes interact with the neutral and ionized atmosphere?

Several observational campaigns were conducted within each project, supported by regular workshops, conference sessions, and business meetings. Results were not only disseminated through the peer-reviewed literature but also to the broader CAWSES community through quarterly TG4 newsletters. This paper describes these activities and reviews the progress made over the CAWSES-II period from 2009 to 2013. Each project and its scientific outcomes are described in a separate section. The manuscript concludes with a general discussion of the outcomes, the scientific challenges for the future, new satellite missions, and SCOSTEP's new VarSITI program. More results related to TG4 can be found in the four special issues listed in Table 2 and in a special issue of Earth, Planets and Space dedicated to results presented at the International CAWSES-II Symposium ([2014]) held at Nagoya University, Japan, 18 to 22 November 2013.

REVIEW

Project 1: How do Atmospheric Waves Connect Tropospheric Weather with Ionosphere/Thermosphere Variability?

The impact of atmospheric waves on ionospheric structure and variability has been realized for quite some time. Hines ([1960]) in his pioneering work was the first to propose GW as the cause of 'irregular motions' in the thermosphere and ionosphere. Since then, numerous studies have revealed that GWs are present at heights up to the upper thermosphere and connected them to medium-scale traveling ionospheric disturbances (MSTIDs), ionospheric irregularities, and plasma instabilities. See for example the more recent review by Fritts and Lund ([2011]) and references therein. Before CAWSES-II, it was already known that GW from convection and jet streams in the lower atmosphere propagate into the mesosphere, dissipate their energy near the mesopause region, and/or penetrate into the thermosphere. However, despite some speculation about the initiation of various plasma instabilities by GW, the relation between GW and MSTID, day-to-day variability and zonal separation of plasma bubbles, and the scale size and propagation of sporadic-E patches were not understood. The examination of the relationship between these phenomena and an improved understanding of the importance of GW in ionosphere/thermosphere dynamics were objectives for CAWSES-II.

The idea that global winds in the ionosphere may be a source of disturbance electric fields and currents goes back at least to the dynamo theory by Stewart ([1882]); the connection to Sun-synchronous (migrating) atmospheric tides forced by solar radiation absorption was discussed by Fejer ([1964]). As tidal theory and observational diagnostics progressed, it was realized that non Sun-synchronous (nonmigrating) tides forced by deep tropical convection are equally important for explaining longitudinal and local time variations in bulk neutral and plasma properties of the ionosphere/thermosphere system. Satellite diagnostics (Forbes et al. [2006]; Oberheide et al.[2006b]; Sagawa et al. [2005]; Immel et al. [2006]) and models (Hagan and Forbes [2002]; Hagan et al. [2007]) resulted in a basic quantitative knowledge of tidal forcing, propagation, and morphology in the mesosphere and lower thermosphere, the ionosphere, and a basic qualitative knowledge about the coupling into the F-region through E-region dynamo modulation. It should be noted that important contributions came from CAWSES-I activities, for example, from the CAWSES tidal campaigns (Ward et al. [2010]) that effectively resolved the long-standing issue between ground-based radar and satellite optical measurements of winds. See also Kishore Kumar et al. ([2013]) for a recent comparison between radar, satellite, and model results obtained during the first CAWSES tidal campaign in 2005. Major challenges for CAWSES-II included the elucidation of tidal structures in the altitude range between 110 (upper altitude observed by TIMED) and 400 km (*in situ* tidal diagnostics from CHAMP) where suitable satellite observations are lacking, temporal variations of the tides on timescales ranging from days to solar cycle, a better separation of E-region dynamo modulation vs tidal coupling at F-region heights, an assessment of the tidal impacts on the energy balance and composition of the thermosphere, and wave-wave and wave-mean flow interactions. Similar questions are applied to PW as well, with special interest in their role in connecting polar stratospheric warmings with F-region low latitude plasma density variability (Goncharenko and Zhang [2008]), as further detailed in project 3.

Gravity Waves

Gravity waves generated by convective systems, hurricanes, and surface perturbations such as earthquakes and tsunamis may seed plasma bubbles in the ionosphere through Rayleigh-Taylor instability. Figure 1 shows GW in airglow images at 630 nm (approximately 250 km) caused by the tsunami from the catastrophic Tohoku earthquake on 11 March 2011 when it passed the Hawaiian Islands. See Makela et al. ([2011]) for details and also for a movie that includes the ionospheric response in total electron content (TEC). The ionospheric response over Japan (Figure 2) for the same event shows very pronounced concentric waves in TEC that propagate in the radial direction with a velocity between 138 and 3,457 m/s with an 'ionospheric epicenter' about 170 km southeast from the earthquake epicenter (Tsugawa et al. [2011]). The initial signatures in Figure 2 with a propagation velocity of 3,457 m/s are most likely caused by a Rayleigh wave (a type of surface wave during earthquakes), followed by acoustic waves triggered by the Rayleigh wave. Finally, medium-scale concentric waves with a propagation velocity of 138 to 423 m/s were detected 300 km away from the ionospheric epicenter (Figure 2g,h). They are the result of GW forced by the surface displacement (Matsumura et al. [2011]).

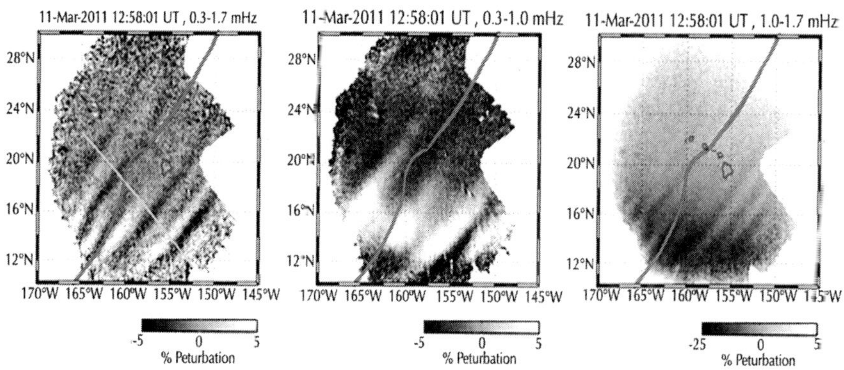

Figure 1: GW in airglow images at 630 nm caused by the tsunami from the Tohoku earthquake. Example of 630.0 nm images processed using length-8 FIR filters with passbands of (left) 0.3 to 1.7 mHz, (middle) 0.3 to 1.0 mHz to highlight the 26.2-min period waves, and (right) 1.0 to 1.7 mHz to highlight the 14.2-min period waves. The red line in each image indicates the tsunami location at the time of the image. The green line in the left panel indicates a

path perpendicular to the airglow wave fronts. Adapted with permission from Makela et al. ([2011]). See also the movie in the electronic supplement to Makela et al. ([2011]).

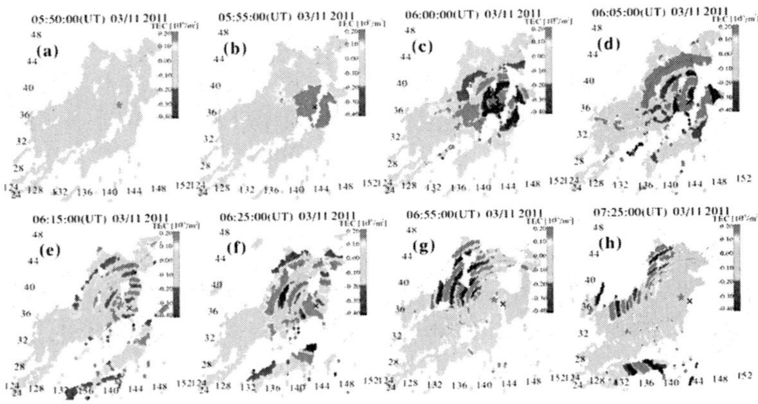

Figure 2: Time series (a)-(h) of detrended TEC during the Tohoku Earthquake on March 11, 2011. The star and cross marks represent the epicenter and the ionospheric epicenter, respectively. Gray circles are about the ionospheric epicenter. Adapted with permission from Tsugawa et al. ([2011]).

Nishioka et al. ([2013]) discussed similar concentric waves in the ionospheric TEC variation over the North-American continent as an indicator of thermospheric gravity waves generated by the 2013 Moore EF5 tornado. Fukushima et al. ([2012]) showed correlation of MSTIDs observed by a 630 nm airglow imager over Indonesia with tropospheric convective activity but with the average horizontal wavelength of the MSTIDs increasing with decreasing solar activity. These findings suggest that the observed MSTIDs in the equator are caused by secondary gravity waves in the thermosphere, possibly generated by tropospheric convective activity. Smith et al. ([2013]) reported on thermospheric secondary GWs generated from mountain wave breaking in the upper mesosphere using a simultaneous observation of 630 nm and 557.7 nm airglow images over New Zealand.

Progress in gravity wave theory (e.g., Vadas and Liu [2009]; Vadas and Crowley [2010]; Yi it et al. [2012]; Gavrilov and Kshetvetskii [2014]) now provides a much clearer picture of how gravity waves from the lower atmosphere can reach thermospheric altitudes, involving a multitude of possible pathways including direct propagation, reflection in the upper atmosphere, and the generation of secondary waves through upper mesospheric and thermospheric body forces (Figure 3). As a result, substantial neutral horizontal wind and plasma density variations from a single convective plume are predicted to occur as far away as adjacent continents, pointing to local wave sources having a global impact: a significant challenge for the interpretation of observations. An example of the ionospheric response to convective activity over Brazil on 1 October 2005 is shown in Figure 4 which is adapted from Vadas and Liu ([2013]). Thermospheric body forces resulting from convective plumes are computed with the Vadas model. Their incorporation into the Thermosphere-Ionosphere-Mesosphere-Electrodynamics General Circulation Model (TIME-GCM) results in large-scale traveling ionospheric disturbances. Of note is that the h_mF2' perturbations are generally anticorrelated with the f_oF2' and TEC' perturbations. This indicates that field-aligned transport instead of chemistry (i.e., [e] + [O^+] recombination) may be the dominant coupling mechanism into the plasma. Observing such signatures still poses a significant challenge. Corresponding temperature perturbations in the thermosphere (not shown) will be observed from geostationary orbit in the near future by the Global-scale Observations of the Limb and Disk (GOLD) instrument (a UV imager to be launched in 2017; Eastes et al. [2008]). For the ionospheric signatures, much progress can be expected from Swarm, an ESA constellation mission (Haagmans et al. [2013]) that was launched in 2013. The role of GW for ionospheric instability seeding was the focus of TG4 project 2 and is further discussed in the corresponding section later in this paper.

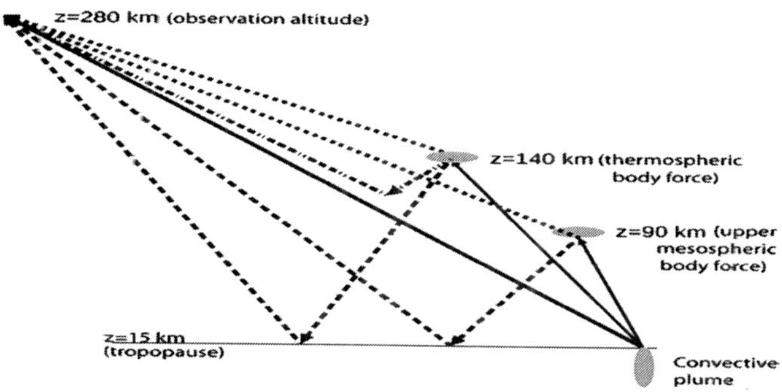

Figure 3: Sketch of possible pathways of a GW. Sketch of possible pathways of a GW from a single convective plume including direct propagation (solid

line), creation of body forces (grey ellipses), and subsequent generation of secondary GW with possible reflection (dotted and dashed-dotted lines). Adapted with permission from Vadas and Crowley ([2010]).

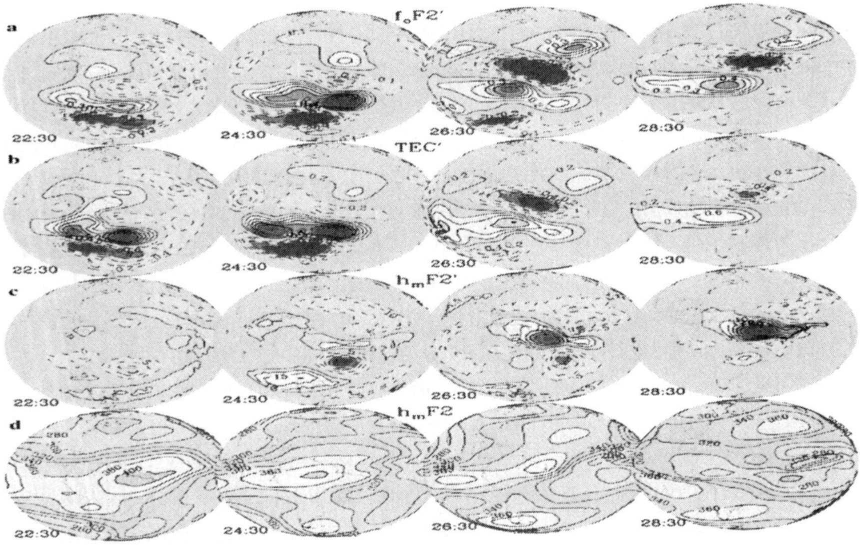

Figure 4: TIME-GCM response to convective activity over Brazil. (a)F2 plasma frequency perturbation (f_oF2' in megahertz), (b) TEC perturbation (in TECU), (c) h_mF2 perturbation (in kilometer), and (d)h_mF2 altitude (in kilometer). These quantities are shown every 2 h from 22:30 to 28:30 UT, as labeled. 1 TECU = 10^{16} electrons m^{-2}. The contour lines are separated by 0.1 MHz, 0.2 TECU, 5 km, and 20 km in panels a to d, respectively. Note that panels a to c show the perturbed minus the unperturbed TIME-GCM solutions, while panel d shows only the perturbed TIME-GCM solution. Adapted with permission from Vadas and Liu ([2013]).

Tides

The components of the tidal spectrum that follow the apparent westward propagation of the Sun relative to the Earth's surface are called migrating tides, and the non-Sun-synchronous components are called nonmigrating tides. Migrating tides are predominantly forced by solar radiation absorption in tropospheric water vapor and stratospheric ozone. Nonmigrating tides are thought to have two major sources:

non-linear wave-wave interaction processes in the strato-/mesosphere (Hagan and Roble [2001]) and latent heating due to large-scale deep convection in the tropical troposphere (Hagan and Forbes [2002], [2003]). Deep convection largely depends on land-sea differences and sea-surface temperatures. Variations in the periodic absorption of solar radiation at the surface thus transform to a longitudinal structure in raindrop formation (heat release) at roughly the same local time of the day that acts as an efficient forcing mechanism for a number of nonmigrating tides including the diurnal eastward propagating tide of zonal wavenumber 3 (DE3). Note that the DE3 appears as a zonal wavenumber 4 (wave-4) longitudinal structure when observed at constant local solar time, e.g., by precessing satellites in a low Earth orbit. This is simply a result of its eastward propagation and frequency.

Figure 5 summarizes the pre-CAWSES-II knowledge of the cause-and-effect chain between latent heat forcing, tidal propagation, and coupling into the ionosphere, based on various satellite observations. The most striking pattern is a 4-peaked 'wave-4' longitudinal modulation that is apparent in deep convective cloud occurrence, E-region zonal winds, thermospheric constituents, F-region electron and ion density, and upper thermosphere neutral mass density. Nonmigrating tidal winds in the low and middle latitude E-region move the partially ionized plasma through Earth's magnetic field while the electrons with their high gyro frequency/collision frequency ratio remain fixed to the magnetic field lines. An electromotive force is thus created with ensuing electric currents and polarization electric fields. The E-region dynamo polarization fields are further transmitted along magnetic field lines into the overlying F-region where they drive vertical (approximately 20 m/s) and horizontal (approximately 100 m/s) plasma drifts, which influence many important ionospheric processes. For example, vertical ExB drifts drive the plasma fountain, which results in dense bands of plasma centered near ±15 to 20 magnetic latitude. This equatorial ionization anomaly can be seen in the top center panel of Figure 5 as the two bluish bands north and south of the equator. The occurrence of the two ionization crests is predominantly the result of migrating solar tidal winds in the E-region, while the apparent longitudinal modulation in these bands is the result of nonmigrating tides excited by latent heat release in the tropical troposphere. For example, electron density varies by a factor of 2 to 3 as a function of longitude along the crests of the equatorial ionization anomaly at ±15 magnetic latitude (top left panel in Figure 5) and reflects the large-scale convective activity in the low-latitude troposphere. This surprising discovery that global conditions in the ionosphere and thermosphere are linked strongly to the terrestrial weather and climate below was initiated by IMAGE satellite observations of the F-region plasma density (Sagawa et al. [2005]; Immel et al. [2006]).

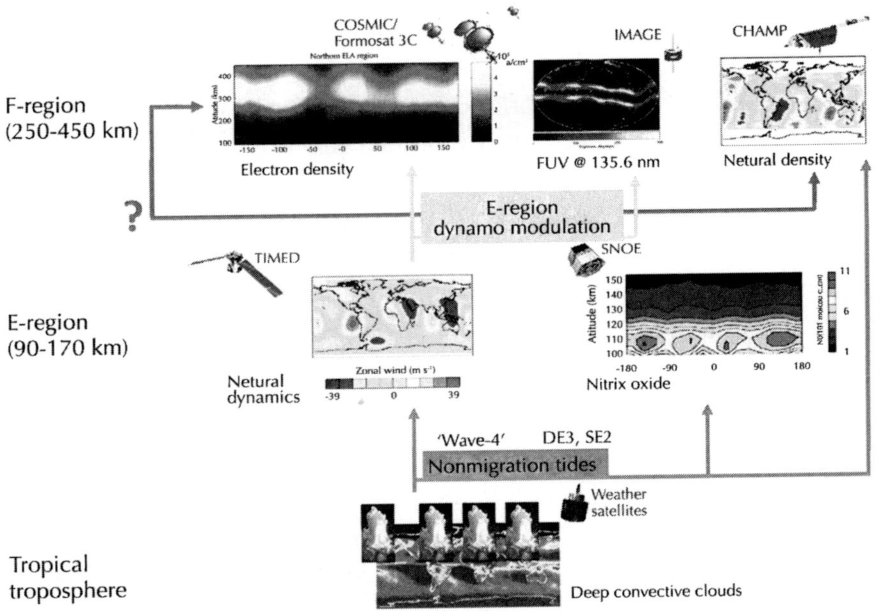

Figure 5: Sketch of meteorological impacts on geospace due to non-Sun-synchronous (nonmigrating) tides. Sketch of meteorological impacts on geospace due to non-Sun-synchronous (nonmigrating) tides as observed by different satellites in various neutral and plasma parameters. The COSMIC figure is adapted with permission from Lin et al. ([2007]), the IMAGE figure is adapted with permission from Immel et al. ([2006]), the TIMED figure uses data from Oberheide et al. ([2011a],[b]), and the SNOE figure is adapted with permission from Oberheide and Forbes ([2008]). Deep convective cloud data are from the ISCCP climatology. CHAMP neutral density data are provided by Dr X. Zhang and Dr. S. L. Bruinsma.

This general picture has been confirmed during the CAWSES-II period by coupled ionosphere/thermosphere models (e.g., Jin et al. [2011]; Maute et al. [2012]) and studies using satellites, e.g., in temperature (Forbes et al. [2009]), infrared cooling (Oberheide et al.

[2013]), and plasma density (Chang et al. [2013]). However, it has now been realized that alternative effects, in addition to the E-region dynamo modulation, may contribute to the coupling between the tides and the ionospheric plasma: for instance neutral density variations, changes in thermospheric atomic oxygen to nitrogen ratio, and meridional winds at F-region altitudes (Liu et al. [2009]; England et al. [2010]; Maute et al. [2012]; see also the review article by England ([2012])).

It should also be noted that tidal wind shears in the E-region play a significant role in forming ionospheric intermediate layers, called sporadic E. See for example the early work by Fujitaka and Tohmatsu ([1973]) and the review by Haldoupis ([2011]). Much progress in this field, particularly in investigating the spatio-temporal distribution of occurrence frequency, has been made during the CAWSES-II period. Important contributions came from the concurrent analysis of COSMIC radio occultation and TIMED tidal diagnostics. For example, it has now been realized that, in addition to the diurnal and semidiurnal tides, the terdiurnal migrating tide plays an appreciable role in the sporadic E formation (Fytterer et al. [2014]). There is also growing evidence for an important role of nonmigrating tides on the sporadic E formation but this topic needs to be studied further in the future.

Some progress was also made in elucidating day-to-day tidal variability. Forbes et al. ([2011]) combined CHAMP and GRACE exosphere temperatures to perform daily tidal fits for selected time periods and found mainly solar-driven variability with a strong correlation with the 27-day solar rotation period (Figure 6). Using combined WACCM-X/TIME-GCM simulations with constant solar and geomagnetic conditions but realistic tropospheric weather patterns, Liu et al. ([2013a], [b]) showed day-to-day tidal amplitude variability on the order of 50% for the DW1, DE2, and DE3 tidal components in thermospheric winds, respectively. This in turn produced ExB vertical drift variability of the same magnitude and clearly indicates that tropospheric weather variability imposes a significant day-to-day variability on the ionosphere. At lower altitudes (mesosphere), Nguyen and Palo ([2013]) combined sounding the atmosphere using broadband emission radiometry (SABER) and microwave limb sounder (MLS) temperature data to derive day-to-day migrating diurnal tide variability at low latitudes and found amplitude changes of up to 15 K from one day to another. This is of the same order as variability deduced from 'deconvolution' approaches (Oberheide et al. [2002])

using SABER alone (Figure 7). The causes for this variability still need to be understood. It is not yet clear what the relative roles of tropospheric source variability and modulations imposed by wave-wave and wave-mean flow interaction processes are in causing this variability. Addressing this challenge is not only a matter of diagnosing short-term tidal variability from the data but will also require a concentrated effort by the modeling community. For example, a comparison of the diurnal tide in models and ground-based observations conducted as part of the 2005 equinox CAWSES tidal campaign (Ward et al. [2010]) points to considerable differences in model magnitudes and vertical wavelengths suggesting that inconsistencies in model forcing, dissipation, and background winds exist (Chang et al. [2012]).

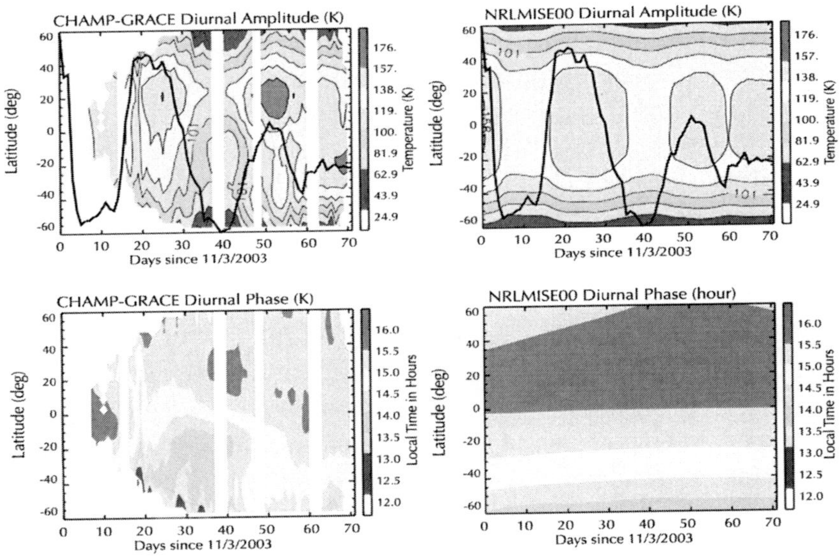

Figure 6: Migrating diurnal tide exosphere temperature amplitudes and phases. Migrating diurnal tide exosphere temperature amplitudes (top) and phases (bottom) from daily CHAMP-GRACE fits (left) and NRLMSISE00 (right) for the period 3 November 2003 to 14 January 2004. The solid line indicates the daily F10.7 solar flux, which ranges from 86 sfu (min) to 190 sfu (max). Adapted with permission from Forbes et al. ([2011]).

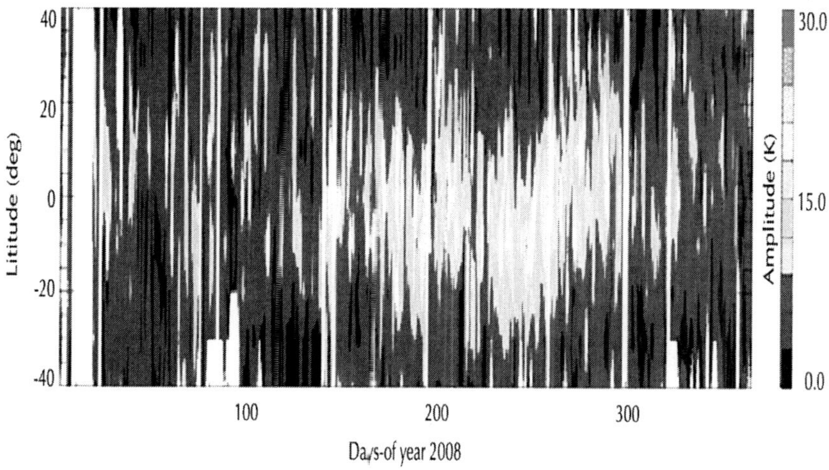

Figure 7: Day-to-day DE3 tidal amplitude variability in 2008 at 100 km from SABER. Day-to-day DE3 tidal amplitude variability in 2008 at 100 km from SABER using a 'deconvolution' approach, plotted as a function of day of year and latitude. White color indicates data gaps.

On longer time-scales, it has become clear that changes in global-scale weather patterns impose a considerable variability in the thermosphere and ionosphere. A compelling example that has recently attracted attention is the effect of the El Niño Southern Oscillation (ENSO). Pedatella and Forbes ([2009a]) found a high correlation between the Oceanic Niño Index (ONI) and longitudinal f_oF_2 variability from low latitude ionosondes and associated this with ENSO-related changes in DE3 convective tidal heating. In a follow-up modeling study using WACCM, Pedatella and Liu ([2012]) showed that ENSO imposes a tidal temperature variability on the order of 10% to 30% during northern hemisphere winter. Interestingly, they found an enhanced DW1 and SW4 during the El Niño phase but a reduced DE2 and DE3. The latter components are enhanced during the La Niña phase. This is consistent with recent TIMED Doppler interferometer (TIDI) diagnostics focused on ENSO variability (Warner and Oberheide [2014]). The ENSO-imposed variability is generally smaller than the short-term tidal variability but similar to variability due to the quasi-biannual oscillation (QBO) determined by Oberheide et al. ([2009]). The ionospheric response to ENSO still needs to be studied in detail. A preliminary analysis (Figure 8) based on the COSMIC-based TEC tides from Chang et al. ([2013]) and the TIDI-, Tropical Rainfall Measuring Mission (TRMM)-, and Modern-Era Retrospective Analysis for Research and Applications (MERRA)-based tidal wind and heating diagnostics from Warner and Oberheide ([2014]) indicates a consistent 50% enhancement in DE3 during the La Niña phase of the strong 2009 to 2011 ENSO and no response during the El Niño phase.

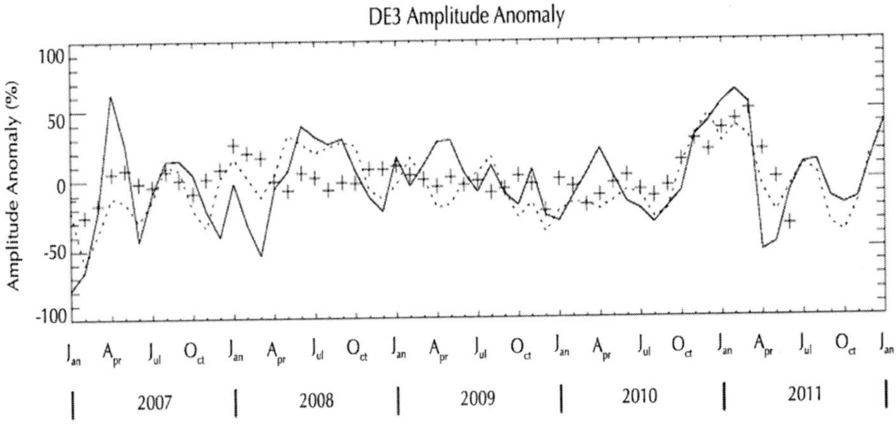

Figure 8: DE3 amplitude anomaly in TEC, E-region zonal wind, and tidal heating. DE3 amplitude anomaly (deviation with respect to the 2007 to 2011 climatological monthly means) in TEC (>200 km) from COSMIC at 15°N magnetic latitude (solid line), TIDI zonal winds at 100 km (averaged between 0°N to 20°N, dotted line), and TRMM- and MERRA-based convective and radiative DE3 heating, averaged from 3 to 16 km (plus symbols). Note the 50% enhancement in winter 2010/11 (La Niña phase) and the lack of any response in winter 2009/10 (El Niño phase).

Tropospheric tidal forcing does not respond appreciatively to the solar cycle, and the E-region dynamo tidal winds remain more or less unaffected (Oberheide et al. [2009]). In the thermosphere above 120 km, however, increasing background temperatures during higher solar activity cause more tidal dissipation because of the temperature dependence of thermal conductivity. For example, DE3 amplitudes during solar minimum are much larger than during solar maximum: a factor of 3 in the zonal wind, 60% in temperature and a factor of 5 in density (Oberheide et al.[2009]; Häusler et al. [2013]). On the other hand, relative TEC tidal amplitudes from COSMIC (Chang et al. [2013]) do not show any solar cycle dependence. This can be understood as the result of the absence of any solar cycle dependence in the zonal wind tides in the dynamo region which are the main driver of ionospheric tidal variability.

Connecting tides observed in the mesosphere lower thermosphere (MLT) region and those in the upper thermosphere and thus making the full connection to tropospheric weather from a purely observational point is still hampered by the lack of global wind or temperature observations in the 'thermospheric gap' between TIMED observations (reliable below 110 km) and *in situ* observations from satellites like CHAMP, GRACE, or the new Swarm mission at altitudes generally above 200 km. For example, much of the tidal characteristics in this height region rely on fits to TIMED observations such as the Climatological Tidal Model of the Thermosphere (CTMT, Oberheide et al.[2011a]) that uses the observed tides in the MLT region as a constraint for a physics-based empirical model or general circulation models like WACCM-X (Liu et al. [2010a]). The only dataset to date that turned out to be suitable for thermospheric tidal diagnostics came from HRDI and WINDII on UARS observations with tidal 'deconvolution' applied (Lieberman et al. [2013a]). Figure 9 exemplifies this for the DE3 using HRDI/WINDII diagnostics from 60 to 250 km compared to CTMT. While CTMT does a reasonable prediction in this example, other components are not well reproduced, particularly DW2 and D0 that have additional *in situ* thermospheric sources, as shown by Jones et al. ([2013]) using the TIME-GCM. The existence of *in situ* thermospheric tidal sources for DW2 and D0 was confirmed by Forbes et al. ([2014]) who used the CTMT and combined CHAMP/GRACE tidal diagnostics to study tidal penetration to the upper thermosphere.

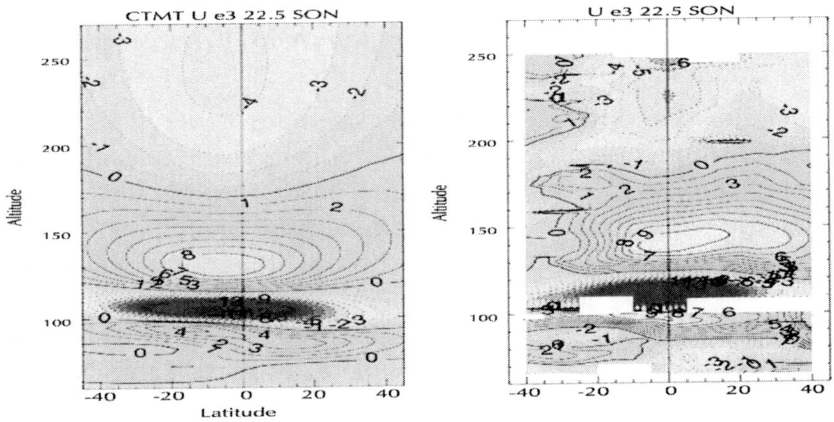

Figure 9: DE3 amplitudes from CTMT and WINDII. (Left) DE3 amplitudes (m/s) computed from CTMT averaged over September to November and mapped to 17:00 to 07:00 local time difference at 22.5°E. (Right) Same as left but from HRDI and WINDII analysis. White areas are data gaps. Adapted with permission from Lieberman et al. (2013).

In continuation of the CAWSES-I tidal campaigns (Ward et al. [2010]), TG4 continued the data analysis (e.g., Chang et al. [2012]; Kishore Kumar et al. [2013]) and conducted an additional campaign in August to October 2011. These campaign data still need to be analyzed and results will be published elsewhere. Although requiring a significant international effort, observational campaigns such as these for various tropospheric and stratospheric conditions (ENSO, QBO) are important for developing and confirming our understanding of tidal variability. See for example the recent modeling results by Gan et al. ([2014]) that highlight interannual variability in satellite-borne tidal diagnostics and in a general circulation model nudged to observed meteorological reanalysis data. It should also be noted that tides provide the link between the 'weather' of the polar stratosphere and ionospheric variability close to the geomagnetic equator, e.g., during sudden stratospheric warmings (SSW) as a result of planetary wave - tidal interaction. This is further elaborated on in project 3.

Project 2: What is The Relation between Atmospheric Waves and Ionospheric Instabilities?

Equatorial plasma instabilities, commonly referred to as equatorial spread-F, plasma bubbles, or depletions can cause radio signals propagating through the disturbed region to scintillate resulting in a distortion or loss of signal. First observed by Booker and Wells ([1938]), they have been extensively studied due to the increasing importance of satellite-based communications and positioning. Much has been learned over the decades about the growth mechanism and occurrence variability on a seasonal timescale. Dungey ([1956]) suggested the Rayleigh-Taylor instability as the generating process during post-sunset at the magnetic equator. At this time and location, since the bottom side of the F layer has recombined while the entire F layer itself has been raised by the pre-reversal enhancement, a very sharp vertical density gradient is created that is unstable to vertical perturbations in the ionosphere. Identifying the contribution of atmospheric waves as a perturbation source and studying the resulting instabilities was the leading motivation for two observational campaigns carried out as part of TG4 activities.

The SpreadFEx-2 Campaign

Based on the success of the first SpreadFEx campaign held in northeastern Brazil (e.g., Fritts et al.[2008]), which focused on studying the interaction between the neutral atmosphere and ionosphere, especially during periods of ionospheric irregularities, a second set of campaigns, called SpreadFEx-2, was carried out in 2009 and 2010. A total of eight institutions participated in these campaigns, deploying instruments to the sites detailed in Table 3. The addition of the coherent electromagnetic radio tomography (CERTO) beacon receivers and deployment of multiple wide-angle imaging systems added the possibility of performing tomographic inversions of the ionosphere during periods of ionospheric irregularities while the deployment of Fabry-Pérot interferometers (FPIs) added the capability to directly measure neutral winds and temperatures and study the relationship between the neutral and plasma state.

Table 2: Special issues with significant TG4 contributions

Title	Journal	Year	TG4 relation
Coupling between the lower and upper atmosphere	JGR Space Physics	2010	General topic
Coupling between the Earth's atmosphere and its plasma environment	Space Science Review, Vol. 168, Issue 1-4, 2012	2012	ISSI workshop, see TG4 newsletter vol. 3
Recent progress in the vertical coupling in the atmosphere-ionosphere system	J. Atmos. Sol. Terr. Phys., Vol. 90-91, pages 1-222, December 2012	2012	4th IAGA/ICMA/CAWSES-II TG4 workshop on vertical coupling
Recent advances in equatorial, low-, and mid-latitude aeronomy	J. Atmos. Sol. Terr. Phys., Vol. 103, pages 1-194, October 2013	2013	ISEA-13 conference

Oberheide et al.

Oberheide et al. Progress in Earth and Planetary Science 2015 2:2 doi:10.1186/s40645-014-0031-4

The campaigns were operated from early September until the end of November in both 2009 and 2010. These months correspond to the beginning of the 'spread-F' season, so called since this is when the occurrence of ionospheric irregularities (known as spread-F or equatorial plasma bubbles) is more common. Combining data from the imaging systems and FPIs, the coupling of the neutrals and plasma during periods of equatorial plasma bubbles (EPBs) was studied. Chapagain et al. ([2012]) found that the zonal neutral winds and EPB zonal drift velocity, which is assumed to be indicative of the background plasma drift velocity, were tightly correlated, especially beginning several hours after sunset. The zonal velocities showed similar patterns of both nightly and night-to-night variability, indicating that the F-region dynamo was fully developed. Earlier in the evening, however, the EPB zonal speed was found to be slower than the background neutral winds, indicating that during the period of bubble development, the F-region dynamo might not be fully activated. Examples of the comparisons made during the SpreadFEx-2 2009 campaign are shown in Figure 10. These examples show the overall similarity between the winds and EPB drift velocities as well as the variability in both of these parameters on a night-to-night basis. Histograms showing the difference between the measured winds and estimated EPB drift velocities are shown in Figure 11 and show the good agreement between the two zonal speeds seen later in the evening (after 23 LT).

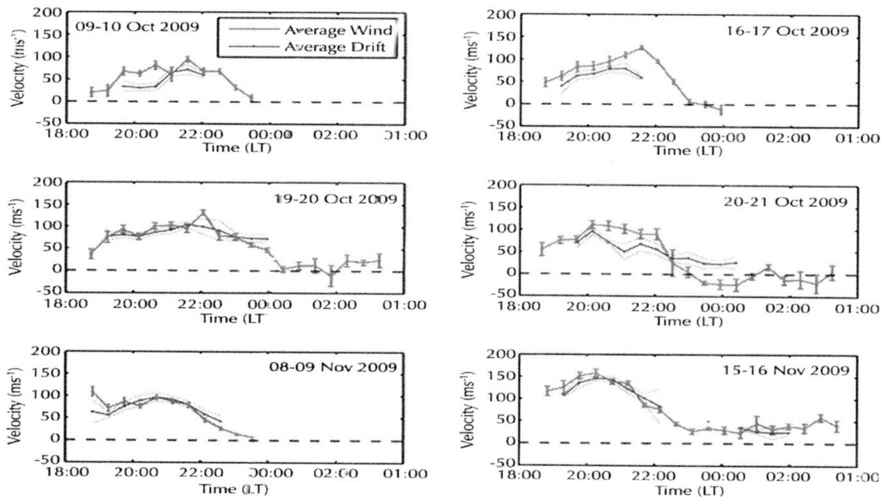

Figure 10: Comparisons of thermospheric neutral winds and equatorial plasma bubble drift velocities. Comparisons of thermospheric neutral winds (red) and equatorial plasma bubble drift velocities (blue) collected during the SpreadFEx-2 campaign in 2009. Adapted with permission from Chapagain et al. ([2012]).

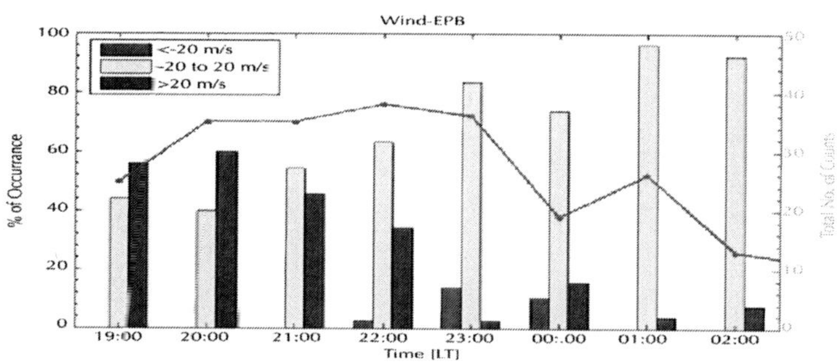

Figure 11: Histogram of differences between the thermospheric neutral wind and equatorial plasma bubble drift velocities. Binned by local time. Total number of coincident observations in each bin is indicated by the blue line, referenced to the right hand axis.

The data collection that begun under the SpreadFEx-2 campaigns has been continued through 2013, allowing for the study of the low latitude thermosphere through the transition from the deep solar minimum toward the solar maximum of solar cycle 24. A summary of the zonal and meridional thermospheric neutral winds as well as thermospheric neutral temperatures approximately 250 km through the end of 2013 are presented in Figure 12. Analysis of the data collected through the end of July 2012 by Makela et al. ([2013]) showed the expected strong dependence of the neutral temperature on the solar flux, with the average temperature increasing by several hundred Kelvin over the study period. However, a solar flux dependence was not seen in the neutral winds, at least over the range of solar fluxes observed during the period. Strong seasonal and day-to-day variability, however, is seen in the data indicating the influence of tidal and, possibly, gravity waves in the thermospheric neutral flows. The collection of this type of multi-year dataset will provide important validation to models of the thermosphere and ionosphere being developed. Such a validation study for the WAM using the dataset that begun under the CAWSES SpreadFEx-2 campaign is presented in Meriwether et al. ([2013]).

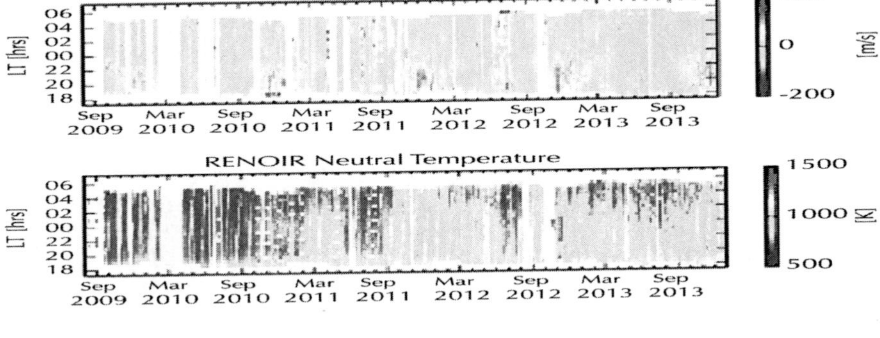

Figure 12: Climatologies of thermospheric neutral winds and neutral temperatures. Climatologies of thermospheric neutral winds (zonal - top; meridional - middle) and neutral temperatures (bottom) collected by the RENOIR FPIs through the end of 2013.

The LONET Campaign

The LONgitudinal NETwork (LONET) campaign was carried out as a TG4 activity during the September to November period in 2010 and 2011 with the goal of creating a robust, global, multi-instrument database to study the longitudinal variability of planetary-scale waves. Table 3 lists the observation sites and instrumentation (13 ionosondes, 1 meteor radar, 2 MF radars, and 1 optical Fabry-Pérot interferometer) participating in LONET. TEC data obtained from the COSMIC satellite were also included. As can be seen from Table 4, the data collected in 2010 were more complete than 2011 in terms of the longitudinal coverage. Figure 13 shows the observation sites distributed along the equator marked on a map. The rectangular boxes depict COSMIC TEC data sampling along the magnetic equator. The ground-based data collection was limited in South America, Asia, and Oceania. No data were available from the African continent.

Table 3: Spread FEx2 observation sites

(Latitude longitude)	Ionosonde	Imager	FPI	Photometer	Meteor radar	VHF radar	CERTO receiver	GPS receiver	3 point sounding	
São Luís	(2.6°S, 44.2°W)	DPS4					OK		OK	
Fortaleza	(3.9°S, 38.4°W)	DPS4								
Cariri	(7.4°S, 36.5°W)	CADI	All sky 6300, OH	6300	OH temperature	OK				Transmission receiver
Cajazeiras			6300	6300				OK	OK	
Petrolina			All sky 6300, OH					OK		
Caico										Transmission
Campina Grande								OK		Transmission.

C. Paulista	(22.7°S, 45.0°W)	DPS4	All sky 6300, OH		Multichannel				OK	
Institution		INPE	INPE, USU, UI	UI and Clemson	UFCG	INPE	INPE	NRL	UI, INPE	UWO, INPE

Oberheide et al.

Oberheide et al. Progress in Earth and Planetary Science 2015 2:2 doi:10.1186/s40645-014-0031-4

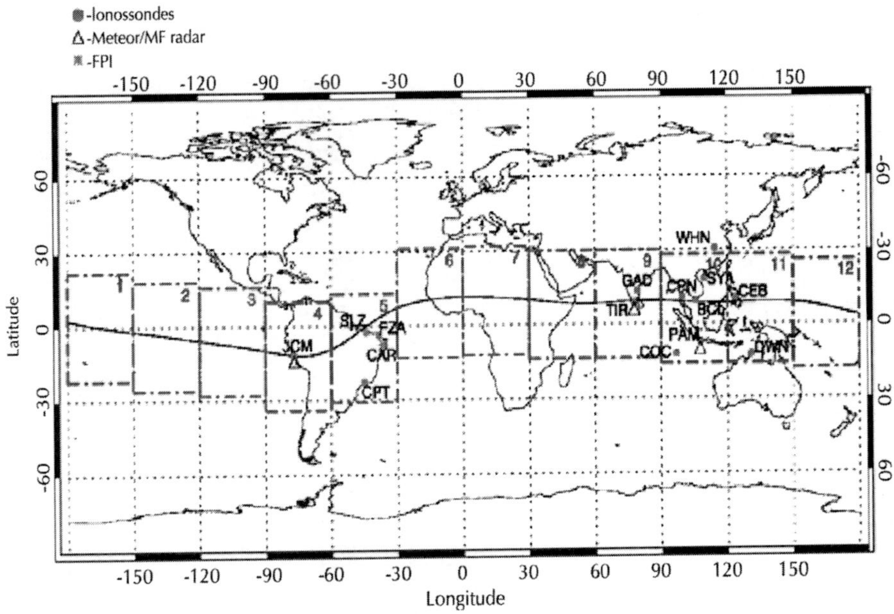

Figure 13: Ground-based observation sites and COSMIC TEC data sampling areas along the geomagnetic equator for LONET campaign.

As the first step in understanding the longitudinal variability of waves, the LONET team studied whether any global-scale periodic oscillations were present in the observed parameters and how the amplitude of those oscillations varied with longitude. From the temporal variations of the ionospheric parameters foF2, h'F, TEC, and mesospheric winds, the 2- to 16-day oscillations were studied using wavelet analysis. In Figure 14, the wavelet power spectra for the three parameters, h'F at Fortaleza (FZA, 321.6°E), TEC at around 345°E, and MF radar zonal wind component at Pameungpeuk (107.7°E) are presented for the period consisting of day of year (DOY) 240 to 330. The h'F and TEC are approximately in the same longitudinal zone. Among the common oscillations seen in these parameters is a distinct 16-day period oscillation during DOY 260 to 300.

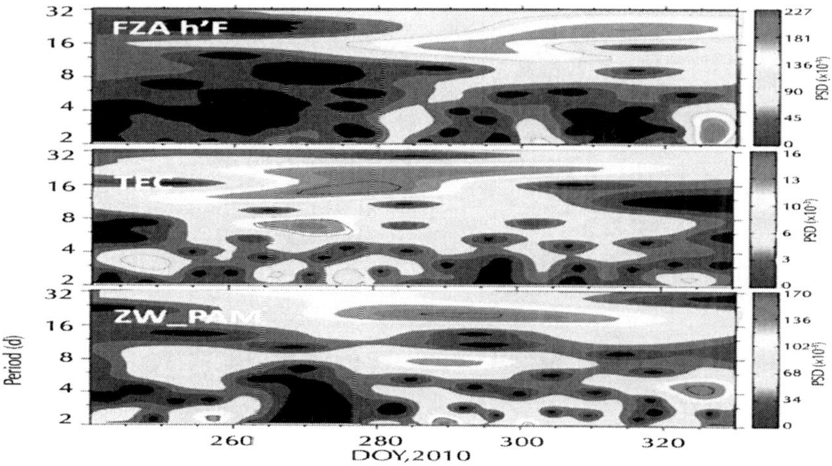

Figure 14: Wavelet spectral density for 2- to 32-day period oscillations during the LONET campaign in 2010. (Top) h'F at Fortaleza, (middle) COSMIC TEC at 345°E, and (bottom) mesospheric zonal wind (89 km height) at Pameungpeuk.

In order to further investigate the 16-day oscillation, the amplitude and phase of the oscillation observed in the h'F parameter observed by the network of ionosondes participating in the LONET campaign are plotted in Figure 15 for Fortaleza (FZA, top) to Darwin (DWN, second from the bottom). The mesospheric wind components (zonal winds at 89 km altitude), obtained from the MF radar at Pameungpeuk (PAM), are also shown (bottom) for reference. From the PAM wind oscillations in Figures 14 and 15, clear evidence of a 16-day oscillation during the period DOY 260 to 320 centered at around DOY 290 is found. The h'F at Fortaleza shows clear 16-day oscillations during the same period with an amplitude that reaches approximately 4 km. However, the amplitudes of oscillation in the Indian and Asian sectors are somewhat lower compared to the American sites.

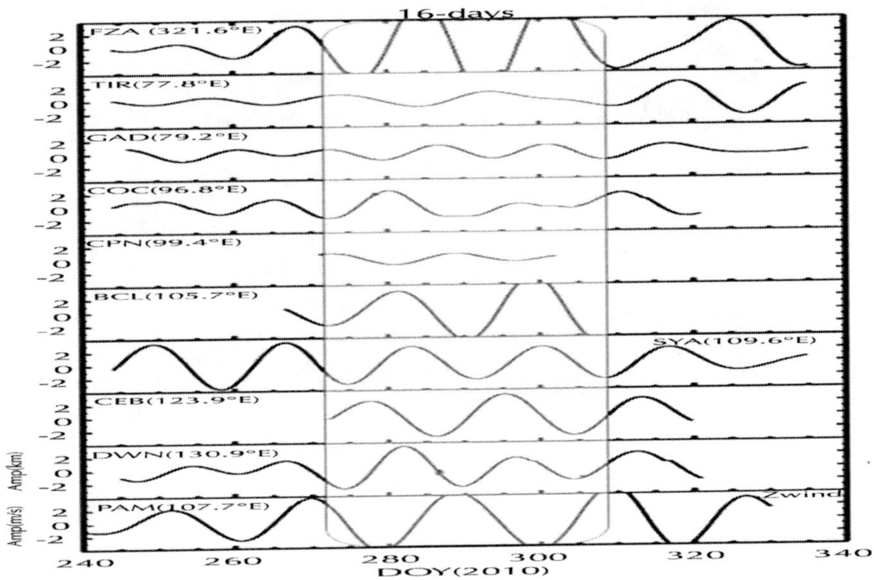

Figure 15: 16-day period oscillations of h'F and mesospheric winds. 16-day period oscillations of h'F and mesospheric winds (bottom) during the LONET campaign period in 2010.

In Figure 16, we plot the results of a 16-day harmonic analysis of the COSMIC TEC data for different longitudinal zones. Each line corresponds to a given longitudinal zone, starting at 195°E (top) to 345°E (middle) to 165°E (bottom). During DOY 270 to 300, the 16-day oscillation feature can clearly be seen. The phase of the maximum is slightly shifted from DOY 290 at 15°E to DOY 300 at 195°E, indicating that the phase is propagating westward. These observations suggest that the 16-day oscillations might be generated by the Rossby mode 16-day planetary wave (Forbes[1995]). Furthermore, there appear to be low amplitude regions (195 to 255°E) and high amplitude regions (315 to 105°E) in the global distribution of the oscillations. The Rossby 16-day waves normally appear in the troposphere to mesosphere mainly through the wind field and propagate into the lower thermosphere, modulating the ionosphere (Lastovicka [2006]). The present results provide further understanding regarding the latitudinal and longitudinal variability of the 16-day waves.

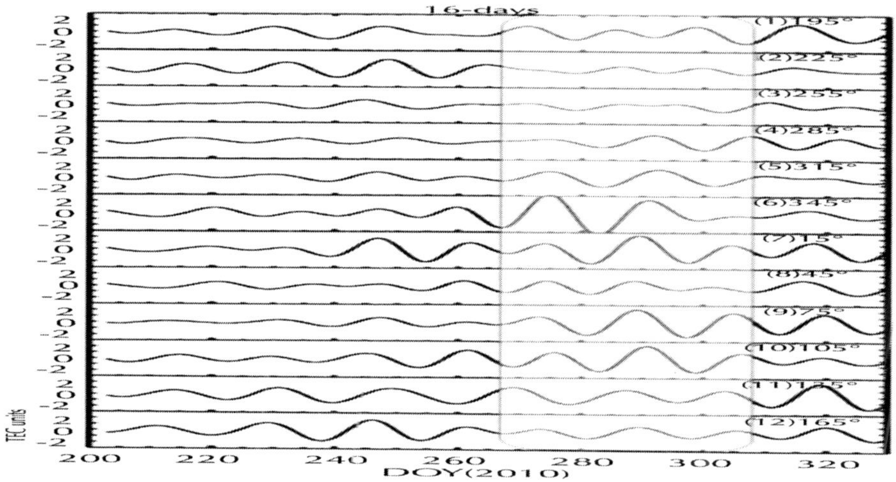

Figure 16: 16-day period oscillations of COSMIC TEC during the LONET campaign period in 2010.

In addition to the ionospheric data presented above, data on geomagnetic field variations were also collected during the campaign period. The results of comparison with the ionospheric parameters will be presented elsewhere. Longitudinal variability of the spread F activity was also investigated using ground-based TEC observation over South America. Large day-to-day variability of spread F activity was observed, and the results were presented in the CAWSES TG4 News letter (vol. 7, page 5, 2012).

Progress of Understanding for Plasma Bubble Seeding By GWs

During the CAWSES-II interval, and in addition to the abovementioned campaigns, several additional interesting results were obtained relevant to the relation between atmospheric waves and ionospheric instabilities. Takahashi et al. ([2009]) showed a positive correlation between plasma bubble spacing and the spatial scale of mesospheric GW, which were observed simultaneously using airglow imagers in 630 nm and OH airglow emissions. Makela et al. ([2010]) found that the distribution of periodic spacing of equatorial plasma bubble compares favorably to the spectrum of GW-induced traveling ionospheric disturbances (TIDs) measured by Vadas and Crowley ([2010]) from a similar geographic latitude in the northern hemisphere. These results suggest that the periodic spacing of plasma bubbles are determined by GWs in the lower thermosphere. Several other studies also suggest seeding of plasma bubbles by GWs in the lower thermosphere (e.g., Taori et al. [2010], [2011]; Paulino et al. [2011]).

Distinct from these small-scale (approximately 100 km) GWs, large-scale wave structures (LSWSs, approximately 1,000 km) have been considered as another controlling factor of the plasma bubble/ equatorial spread F (ESF) generation (e.g., Tsunoda [2005]). Thampi et al. ([2009]) and Tsunoda et al. ([2010]) demonstrated that a close relationship exists between LSWS and the generation of ESF when the post-sunset rise (PSSR) of the F layer was absent. Tsunoda et al. ([2011]) showed that the amplification of LSWS in the late afternoon mostly occurs during the post-sunset rise of the equatorial F layer. Narayanan et al. ([2014]) showed that the occurrences of ionogram satellite traces, which are used as a proxy of LSWS, were followed by ESF in about 71% of the cases, supporting the view that LSWS appears to be an important parameter in the formation of ESF.

It is also interesting to note that Otsuka et al. ([2012]) and Shiokawa et al. ([2014]) reported observations of the dissipation of plasma bubbles in the field-of-view of 630 nm airglow images due to collision of the bubbles with a mid-latitude MSTID and a large-scale TAD, respectively. In both cases, the bubbles seem to be dissipated by polarization electric field associated with these waves in the ionosphere and thermosphere. This process suggests an additional role of GWs in the ionosphere, namely the suppression of ionospheric instabilities.

Project 3: How do the Different Types of Waves Interact as They Propagate through the Stratosphere to the Ionosphere?

A variety of satellite- and ground-based data sets along with model simulations have been used during the CAWSES-II time frame to delineate how large-scale waves propagate through the middle atmosphere to the ionosphere. Several dedicated campaigns, e.g., coordinated ISR observations during SSWs, and workshops (Table 2) involved the TG4 community and were supported by TG4.

Table 4: LONET campaign observation sites

Item	Site	Symbol	Latitude	Longitude	2010	2011	Contact
Ionosonde/foF2, h'F							
1	Jicamarca	JCM	12.0°S	76.9°W	Partial	YES	Karim Kuyeng/ Chau
2	Fortaleza	FZA	3.9°S	38.4°W	YES	YES	Inez Batista
3	C. Paulista	CPT	22.7°S	45.0°W	YES	YES	Inez Batista
4	São Luís	SLS	2.5°S	44.3°W	YES	YES	Inez Batista
5	Tirunelveli	TIR	8.7°N	77.8°E	YES	NO	Diwarker/ Gurubaran
6	Gadanki	GAD	13.5°N	79.2°E	YES	NO	A.K. Patra
7	Cocos Islands	COC	12.2°S	96.8°E	YES	YES	Phil Wilkinson
8	Chumphon	CPN	10.7°N	99.4°E	YES	NO	Nagatsuma
9	Bac Lieu	BCL	9.3°N	105.7°E	YES	NO	Nagatsuma
10	Sanya	SYA	18.3°N	109.6°E	YES	YES	Guozhu Li
11	Wuhan	WHN	30.5°N	114.3°E	YES	YES	Guozhu Li
12	Cebu	CEB	10.3°N	123.9°E	YES	NO	Nagatsuma
13	Darwin	DWN	12.5°S	130.9°E	YES	YES	Phil Wilkinson
Meteor radar/wind							
	Jicamarca	JCM	12.0°S	76.9°W	Partial	YES	Luiz/Chau
MF radar/wind							
1	Tirunelveli	TIR	8.7°N	77.8°E	Partial	YES	Narayanan/ Gurubaran
2	Pameungpeuk	PAM	7.6°S	107.7°E	YES	YES	T. Tsuda/ IUGONET
FPI (thermospheric winds)							
	Cairi	CAR	7.5°S	36.4°W	YES	YES	Makela/ Meriwether
COSMIC satellite data (TEC)							
	Equatorial		Geomag. ±20		YES	YES	COSMIC

Oberheide et al.

Oberheide et al. Progress in Earth and Planetary Science 2015 2:2 doi:10.1186/s40645-014-0031-4

PW Signatures in the Ionosphere

For example, the global distribution and the climatological features of the 5- to 6-day planetary waves derived from SABER/TIMED temperature measurements over the height range 20 to 120 km for a full 6-year period were delineated by Pancheva et al. ([2010]). This study determined that the 5-day wave, the gravest symmetric wave number 1 Rossby wave, has a vertically propagating phase structure with a mean vertical wavelength of approximately 50 to 60 km, whereas the 6-day wave is an equatorially trapped wave number 1 eastward propagating wave with a mean vertical wavelength of approximately 25 km. Liu et al. ([2010a],[b]) attributed temporal hmF2 'wave-4' variations diagnosed from COSMIC data to the interaction of the DE3 with the 5- and 2-day planetary waves. In another study, Mukhtarov et al. ([2010]) reported on the climatology of stationary planetary waves with zonal wave numbers 1 and 2 observed in SABER/TIMED temperatures for the same period. Further, using COSMIC data on ionospheric parameters, f_oF_2 and hmF2 and electron density at fixed altitudes and the SABER temperature data, Pancheva and Mukhtarov ([2012]) diagnosed the global-scale spatial and temporal variability of the approximately 23-day zonally symmetric planetary wave during the Northern winter of 2008/2009. The ionospheric oscillations were shown to be forced by planetary waves propagating from below. Pedatella and Forbes ([2009b]) demonstrated the presence of 16-day waves that were simultaneously observed in the CHAMP electron densities at 350 km, total electron content from global positioning system (GPS) observations, and in SABER temperature data at altitudes of 110 km, indicating that vertically propagating planetary waves induce significant variability in the low latitude F-region ionosphere and that these effects are seen at all longitudes. Recently, Chang et al. ([2013]) analyzed COSMIC TEC data and found strong stationary planetary wavenumber 4 (SPW4) signals (Figure 17), most likely caused by a strong SPW4 in neutral zonal wind in the E-region (Oberheide et al. [2011b]) that in turn is the result of a modulation of the migrating diurnal tide DW1 by the DE3 nonmigrating tide (Hagan et al. [2009]).

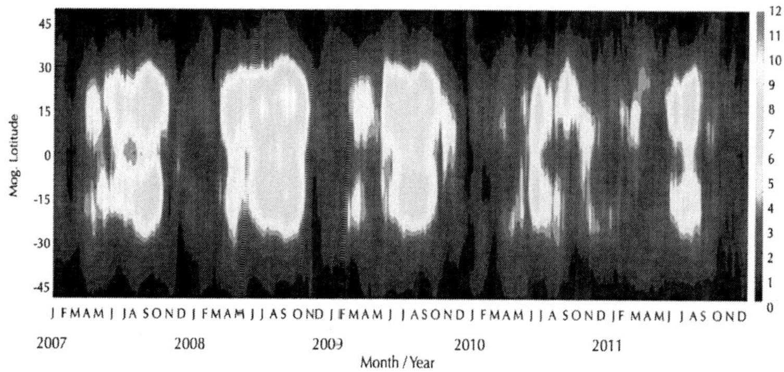

Figure 17: Relative SPW4 TEC amplitude in percent from COSMIC as a function of latitude and year. TEC was computed from the observed electron densities above 200 km. Adapted with permission from Chang et al. (2013).

While the presence of PW signatures in the ionospheric plasma is without dispute and the same neutral/plasma coupling processes as for the tides (predominantly E-region dynamo modulation, see project 1) are thought to apply to PWs as well, the mechanism through which they enter the E-region is still under debate. For example, planetary waves do not penetrate much above 100 km, but instead are thought to impose their periodicities on the ionosphere and thermosphere by modulating the tides and gravity waves that do penetrate to higher altitudes. Other mechanisms include secondary PW forcing. For instance, Lieberman et al. ([2013b]) using MLS, SABER, and TIDI data, along with Navy Operational Global Atmospheric Prediction System - Advanced Level Physics-High Altitude (NOGAPS-ALPHA) model simulations found observational evidence for wintertime SPWs in the lower thermosphere forced in part by drag imparted by gravity waves that have been modulated by underlying stratospheric SPWs. Their results supported earlier model and case studies by Liu and Roble ([2002]), Smith ([2003]), and Oberheide et al. ([2006a]) that already suggested the plausibility of this mechanism.

SSW Impact on the Ionosphere and Thermosphere

A prime example for tidal modulation by PWs occurs during sudden stratospheric warmings. An SSW is an extreme atmospheric disturbance event that results in a warmer polar stratosphere and reversed zonal mean flow at mid-latitudes and provides the link between the 'weather' of the polar stratosphere and ionospheric variability close to the geomagnetic equator. As a consequence of publication of the observational evidence for the ionospheric signatures associated with a SSW event by Goncharenko and Zhang ([2008]), several studies in the recent past have addressed the effect of SSW on the ionosphere due to planetary wave propagation from below. It is now understood that the planetary waves modulate the propagation characteristics of predominantly the semidiurnal migrating tide, enhancing its amplitude, and thereby modulating the E-region ionospheric dynamo, vertical plasma drifts, and plasma density in the F-region, in a manner similar to the effects of nonmigrating tides (Liu et al. [2010b]; Fuller-Rowell et al. [2011]; Chau et al.[2012]; Jin et al. [2012]). This is exemplified in Figure 18 in which GAIA model simulations are compared with COSMIC and SABER satellite observations during the January 2009 SSW.

Figure 18: Comparison of the amplitude of the SW2 semidiurnal migrating tidal component. Comparison of the amplitude of the SW2 semidiurnal migrating tidal component at different altitudes extracted from GAIA model simulations (left) and from COSMIC and SABER satellite observations (right). Adapted from the Jin et al. TG4 Newsletter 13 (2013) article using results from Jin et al. ([2012]).

A recent ground-based study by Laskar et al. ([2014]) conducted in the Indian sector suggests that the 16-day modulation of the semidiurnal tide is particularly strong during strong SSW events and as such exerts a significant impact on ionospheric variability even during high solar activity. Although SSWs only happen at most a few times per year, they nevertheless provide an excellent opportunity to test current theories of neutral-plasma coupling on a global scale and for testing the predictive capabilities of space weather models, as SSWs can be forecasted on a 1-week timescale (Fuller Rowell et al. [2011]; Wang et al. [2014]).

Pancheva and Mukhtarov ([2011]) presented for the first time the global spatial (latitude and altitude) structure of the mean ionospheric response to SSW events during winters of 2007/2008 and 2008/2009 using COSMIC f_oF2 and h_mF2 and electron density data at fixed altitudes. Several studies were carried out to understand the processes underlining the SSW control of the electrodynamics at low latitudes. Chau et al. ([2009]) provided strong evidence indicating that the distinctive behavior of the vertical ExB drifts over the magnetic equator during daytime was associated with a minor SSW. Following this work, Chau et al. ([2010]) identified SSW signatures in several ionospheric parameters using the incoherent scatter radar (ISR) electron density and temperature measurements from the Arecibo Observatory, as well as relative TEC variations derived from a dual-frequency GPS receiver. To examine the response of the mid-latitude ionosphere to SSW, Goncharenko et al. ([2013]) undertook a case study of the day-to-day variability in the ion temperature at altitudes between 200 and 400 km and detected disturbances at tidal periods as well as at non-tidal and multi-day periods. As planetary waves are not expected to reach middle and upper thermospheric altitudes, as noted in the previous subsection, these results have once again raised questions about the underlying mechanisms coupling the lower and upper atmosphere.

A TIME-GCM simulation by Liu and Roble ([2002]) revealed that the resonant SPW amplification prior to the peak warming causes a deceleration of the mean wind in the high latitude winter stratopause and mesosphere and reverses to westward with a critical layer near the zero wind line. This changes the filtering of gravity waves by allowing more eastward GW to propagate into the MLT region and a resulting change of the meridional circulation in the upper mesosphere from poleward/downward to equatorward/upward with corresponding stratospheric warming, mesospheric cooling, and thermospheric warming patterns in the mean temperatures at high latitudes during SSWs. At low latitudes, this temperature pattern reverses and produces a thermospheric cooling and consequently a thermospheric density decrease at a fixed altitude by reducing the scale height.

SSW and Lunar Tides

Recent ground-based and satellite magnetometer observations in conjunction with other ionospheric measurements have provided insights of how lunar tidal modulation during SSW events can drive the spatio-temporal variabilities of the equatorial electrojet and the electrodynamics (Fejer et al. [2010], Park et al. [2012], Yamazaki et al. [2012]). However, Stening ([2011]) argued against a strong lunar tidal contribution to the observed variability during SSWs since strong lunar tidal signals exist during non-SSW periods so that these correlations might be coincidental. Fuller-Rowell et al. ([2011]) indeed found SSW-induced phase shifts in the semidiurnal solar tide consistent with an apparent lunar tidal signal using WAM simulations, but it should be noted that the model did not include the semidiurnal lunar tide as an input. Wang et al. ([2014]) in a coupled WAM/ionosphere model simulation also found a phase shift of the SW2 solar component during SSWs caused by SSW-induced variability in its main source, stratospheric ozone, and/or additional middle atmosphere circulation changes. On the other hand, Pedatella et al. ([2014]) reported notable enhancements of the semidiurnal lunar tide during SSW events in WACCM-driven TIME-GCM simulations and a better agreement with COSMIC electron density observations when the lunar tide was included in the model. It must thus be concluded that the impact of lunar tides on ionospheric variability during SSWs is not yet finally resolved and requires further studies by the aeronomy community over the coming years. It should also be noted that there is increasing evidence that the ionospheric plasma variability during SSWs is not solely caused by E-region electric field variability but also by variability in F-region meridional neutral winds and thermospheric composition changes (e.g., Pedatella et al. [2014]). This is consistent with the earlier modeling work by England et al. ([2010]) and Maute et al. ([2012]) that was focused on nonmigrating tides.

SSW and GW

Small-scale variability of the upper atmosphere driven by major atmospheric disturbances from below like SSW has also gathered attention recently. Adopting a first principle nonlinear hydrostatic GCM extending from the lower atmosphere to the thermosphere, Yi it et al. ([2014]) investigated the influence of small-scale gravity waves originating in the lower atmosphere on the variability of the high latitude thermosphere during an SSW event. The numerical experiments revealed that the gravity wave penetration into the thermosphere increased the momentum deposition rates above 150 km in the high latitude northern hemisphere by up to a factor of 3 to 6 during the warming. This demonstrates that gravity wave-induced variations during SSWs constitute a significant source of high latitude thermospheric variability.

Project 4: How do Thermospheric Disturbances Generated by Auroral Processes Interact with the Neutral and Ionized Atmosphere?

Upper Thermosphere

From the CHAMP satellite observations, Lühr et al. ([2004]) showed the existence of strong heating in the cusp (and/or near the cusp) region. This heating causes upwelling of the air and enhancement of the neutral mass density in the altitude region of about 400 km. In addition, the CHAMP satellite observed some thermospheric disturbances during geomagnetically disturbed periods. For example, Ritter et al. ([2010]) identified substorm-related thermospheric density and wind disturbances from the CHAMP observations. They reported mass density enhancements, a density bulge propagating as a traveling atmospheric disturbance (TAD), and wind variations during a substorm event. Liu and Yamamoto ([2011]) reported geomagnetic storm effects on the formation and characteristics of the mid-latitude summer nighttime anomaly (MSNA), which is a phenomenon during which the diurnal variation of the plasma density maximizes at night instead of day. They pointed out the role of the effective neutral wind in the formation of MSNA. Some results obtained from the CHAMP observations were summarized in the review paper presented by Lühr et al. ([2012]).

Vickers et al. ([2013]) developed a new technique to obtain thermospheric neutral density at approximately 350 km altitude from incoherent scatter radar data. *In situ* comparisons with the CHAMP satellite show good agreement. Vickers et al. ([2014]) used this technique on the EISCAT Svalbard radar for the period 2000 to 2013 and showed that F10.7 solar irradiance is a very good proxy for thermospheric density variations with only a small seasonal dependence. In addition, the long-term trend of declining thermospheric density was observed.

Co-rotating interaction region (CIR) and subsequent high speed solar wind stream (HSS) is one of important causes of the thermospheric disturbances in association with auroral phenomena during a solar minimum period (e.g., Thayer et al. [2008]; Lei et al. [2011]; Verbanac et al. [2011]). Pedatella and Forbes ([2011]) described responses of the ionosphere and thermosphere to high speed solar wind streams from the DMSP F13 satellite observations and the Thermosphere-Ionosphere-Electrodynamics General Circulation Model (TIE-GCM) simulations. They concluded that the disturbance dynamo would be an important mechanism for driving the electrodynamic response at dawn and dusk to recurrent geomagnetic activity driven by HSS. Gardner et al. ([2012]) investigated the HSS heating effects from numerical simulations and observations with the EISCAT Svalbard radar (ESR) and PFISR. They showed that the HSS heating can cause temperature enhancement as high as 100 K at high latitudes, but the global increase in thermospheric temperature would be low.

Coronal mass ejections (CMEs) are also one of the important causes of the geomagnetic storms. The thermospheric density variations during CME- and CIR-induced geomagnetic activities were compared by Chen et al. ([2012]). The total changes in the thermospheric density observed during periods of CIR storms were greater than those of the CME storms because the CIR storms lasted longer than CME storms, while the CME storms were stronger than the CIR storms on average.

For smaller scale waves, several airglow imagers investigated thermospheric and ionospheric waves identified as MSTIDs near the auroral zone through the 630 nm airglow emissions at altitudes of 200 to 300 km. Kubota et al. ([2011]) reported characteristics of MSTIDs observed by an airglow imager at Alaska. They concluded that these southwestward-moving MSTIDs are not caused by ionospheric instabilities, as usually suggested at middle latitudes, but are likely to be caused by auroral disturbances as TADs because the observed background thermospheric wind was poleward, stabilizing the ionospheric instability. On the other hand, Shiokawa et al. ([2013]) reported similar southwestward-moving MSTIDs over Norway and northern Canada and suggested that they are caused mainly by the Perkins and E-F coupling instabilities (Perkins [1973]; Yokoyama et al. [2009]) similar to those at middle latitudes and that an additional source by atmospheric gravity waves from lower altitudes also comes into play. Shiokawa et al. ([2012],[2013]) reported sudden movements of these MSTIDs at subauroral latitudes concurrent with auroral activity, indicating instantaneous penetration of auroral electric fields to subauroral latitudes.

MLT Region

Dynamics and Energetics of the Lower Thermosphere in Aurora (DELTA) and DELTA-2 campaigns were carried out in December 2004 and January 2009, respectively. Simultaneous observations with the EISCAT radar, FPI, and sounding rockets were successfully made during the two campaigns. For example, Kurihara et al. ([2009]) identified temperature enhancement in the lower thermosphere in association with the auroral heating during the DELTA campaign. Oyama et al. ([2010]) also reported lower thermospheric wind fluctuations measured with an FPI during pulsating aurora at Tromsø, Norway, in the period of the DELTA-2 campaign (Figure 19). Some studies concerning the occurrence of atmospheric gravity waves in the polar thermosphere in response to auroral activity were summarized in the review paper of Oyama and Watkins ([2012]).

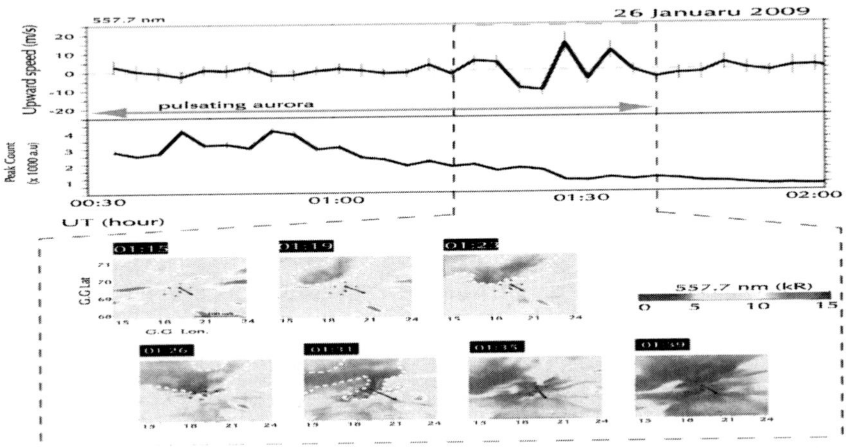

Figure 19: Temporal variations in the vertical wind speed and the fringe peak count at the zenith. Top and middle panels show the temporal variations in the vertical wind speed and the fringe peak count at the zenith, respectively, measured with the FPI (557.7 nm) from 00:30 to 02:00 UT on 26 January 2009. Vertical bars are the 2 sigma uncertainty (see in the text of Oyama et al. [2010]). The bottom color panels show the horizontal aurora images taken with the ASC at 557.7 nm from 01:15 to 01:39 UT during pulsating aurora (corresponding time interval is marked by black dashed line in the top two panels). The color scale of optical intensity is shown at the right middle. These images are mapped in geographic coordinate assuming the peak emission height is 110 km. The five black dots and the four red crosses indicate the location of the FPI and the EISCAT UHF radar observation positions, respectively. The black arrows correspond to the horizontal component of the FPI-derived neutral wind velocity (the scale is presented at the right bottom corner of the first panel from the left). White dashed lines in the colored panels are drawn to easily identify the darker area. Adapted with permission under Creative Commons Attribution 3.0 License from Oyama et al. ([2010]).

Following initial studies of thermospheric meso-scale variability using tristatic Fabry-Pérot interferometer (FPI) observations at EISCAT (Aruliah et al. [2004], [2005]), the neutral wind variations in the vicinity of an auroral arc were studied by some researchers. Oyama et al. ([2009]) showed spatial evolution of frictional heating and increases in ion flow and temperature in the vicinity of an auroral arc from observations with the Sondrestrom incoherent-scatter radar and the Reimei satellite. They clarified localized ionospheric structures associated with localized soft particle precipitation or F-region ionization, suggesting the presence of a narrow thermospheric wind shear of about 10 km width. Kosch et al. ([2010]) reported the first observational results of the E-region neutral wind fields and their interaction with auroral arcs during geomagnetically quiet conditions at Mawson, Antarctica, with a scanning Doppler imager (SDI). The 144° field-of-view Doppler images of the sky over Mawson are shown in Figure 20. They found that E-region wind could rotate 90° within approximately 10 min when close to (within approximately 50 km) of an auroral arc and then recover to the background flow when the arc disappeared. The F-region wind remained unaffected. This was attributed to the electric field associated with auroral arcs causing localized strong ion drag. Anderson et al. ([2011]) performed bistatic FPI observations of F-region thermospheric vertical winds in Antarctica. They found strong upward winds poleward of the auroras as well as correlated vertical wind responses over horizontal distances of approximately 150 to 480 km.

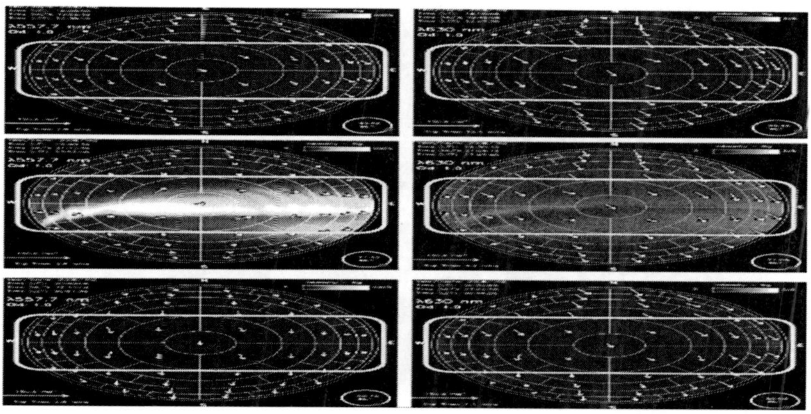

Figure 20: The 144° field-of-view Doppler images of the sky over Mawson. The right column shows F layer 630 nm Doppler images, the left column shows E layer 557.7 nm Doppler images on 10 June 2008. Each plot is labeled with UT. The optical intensity is not calibrated. The optical data are plotted in geographic coordinates with horizontal wind vectors overlaid for each zone (yellow). The wind vectors are centered on their measurement points. The vector for the zenith sector shows the average wind field, not the vertical wind. The top row (18:16 to 18:27 UT) shows before the arc appeared, the middle row (19:36 to 19:40 UT) shows when the arc appeared, and the bottom row (20:33-20:40 UT) shows after the arc disappeared. The blue box highlights the region of interest.

Kosch et al. ([2011]) made meso-scale observations of the ionosphere and thermosphere with the EISCAT Svalbard radar and SDI, partly in the vicinity of an auroral arc, within the polar cap at approximately 110 km altitude. They found that Joule heating had a meso-scale (approximately 64 km) spatial variability of up to >100% even during steady-state geomagnetic conditions. Figure 21 shows the spatial distribution of Joule heating in 22.5° sectors about the EISCAT Svalbard radar projected to 110 km for 23:36 to 23:54 UT on 2 February 2010. They estimated the E-region ion-neutral collision frequency at 112.8 (±0.8) km altitude by combining E- and F-region ion flow with E-region neutral flow data and found good agreement with the MSIS model. Anderson et al. ([2013]) also made observations of the ionosphere and thermosphere with PFISR and SDI. They found the neutral wind dynamo contribution to Joule heating was 36% to 64% on average and was positive when the horizontal magnetic field perturbation was weak (| H| < approximately 40 nT) and vice versa. They estimated the F-region ion - neutral collision frequency at 260 km altitude using the ambipolar diffusion equation - and also found good agreement with the MSIS model.

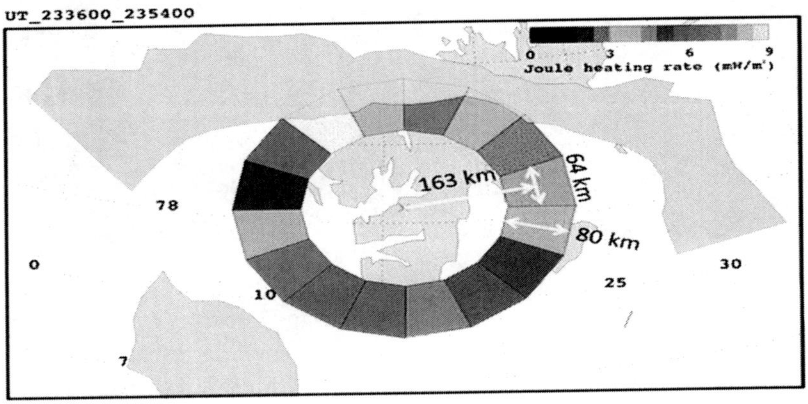

Figure 21: The spatial distribution of Joule heating in 22.5° sectors. The spatial distribution of Joule heating in 22.5° sectors about the EISCAT Svalbard radar projected to 110 km 23:36 to 23:54 UT on 2 February 2010. The aurora (areas in green color) is overlaid for 110 km altitude.

Recently, Xu et al. ([2013]) showed longitudinal variations of the thermospheric temperature from satellite observations with SABER and

Michelson Interferometer for Passive Atmospheric Sounding (MIPAS) and simulations with TIE-GCM. Their simulations clarified that auroral heating would cause the observed longitudinal temperature variations. In addition, the satellite observations indicated that impacts of auroral heating on the neutral atmosphere can penetrate down to about 105 km. The lidar observations are quite useful to study thermospheric disturbances in the MLT region. For example, Chu et al. ([2011]) observed temperature enhancements caused by Joule heating and fast gravity waves propagating from the lower atmosphere in the neutral Fe layers (110 to 155 km) in Antarctica. Tsuda et al. ([2013]) showed decreases in sodium density in association with auroral particle precipitation from simultaneous and common volume observations by the EISCAT VHF radar and a sodium lidar at Tromsø, Norway.

Model Simulations in Association with Auroral/ High Latitude Phenomena

Some general circulation models (GCMs) have been developed to understand the energetics, dynamics, and composition variations of the upper atmosphere. These simulations clarified disturbances in the polar ionosphere and thermosphere in association with auroral activity and/or high latitude energy inputs. For example, Qian et al. ([2010]) simulated thermospheric response to recurrent geomagnetic forcing with the NCAR TIE-GCM. Neutral density and nitric oxide (NO) cooling rates were simulated for the declining phase of solar cycle 23. The simulated results were compared to neutral density derived from satellite drag and to NO cooling measured by TIMED/SABER. Gardner and Schunk ([2011]) performed numerical simulations to investigate characteristics of the large-scale gravity waves with a high-resolution global thermosphere-ionosphere model which can represent global distributions of mass density, temperature, and all three components of the neutral wind at altitudes from 90 to 500 km without the assumption of hydrostatic equilibrium. They noted that a secondary wave was generated which propagated around the entire globe, while the original gravity wave was localized. Deng et al. ([2011]) also investigated nonhydrostatic effects in the thermosphere from Global Ionosphere Thermosphere Model (GITM). Their simulations demonstrated that most of the nonhydrostatic effects at high altitudes (300 km) arise from sources below 150 km and propagated vertically through the acoustic wave. Basic structures of the polar ionosphere and thermosphere during solar minimum and geomagnetically quiet periods were investigated by Fujiwara et al. ([2012]) from the EISCAT Svalbard radar observations (March 2007 to February 2008) and simulations with a whole atmosphere GCM. Their results indicated that both the ions and neutrals would show larger variations than those described by the empirical models, suggesting significant heat sources in the polar cap region even under solar minimum and geomagnetically quiet conditions. In addition to the global model simulations, de Larquier et al. ([2010]) performed one- and two-dimensional numerical simulations with a finite-difference time-domain (FDTD) model to provide quantitative interpretation of the recently reported infrasound signatures from pulsating aurora. They discussed pressure perturbations caused by particle heating due to pulsating aurora (heat source is located between 90 and 110 km on average). The simulation results were roughly an order of magnitude smaller than those observed, suggesting the need for an additional source, e.g., Joule heating.

CONCLUSIONS

Scientific Outcome and Challenges

Overall, TG4 brought together the historically separate neutral atmosphere and plasma communities in a way that allowed for much progress in understanding how neutral variability originating in the lower atmosphere impacts and interacts with Earth's ionosphere, from low to polar latitudes and from the troposphere to the F-region. For example, the role of waves due to convection, polar stratosphere dynamics, or auroral processes in causing substantial ionospheric variability through dynamo processes and Rayleigh-Taylor instabilities is now much better understood than before. This was achieved through a combination of dedicated campaign activities and workshops resulting in four special issues in the peer-reviewed literature (Table 2). The aeronomy community is well on its way to separating ionospheric variability introduced by driving from below and from the magnetosphere and Sun above, an essential task toward achieving predictability of space weather. In this respect, it is important to note that the solar activity at the beginning of the CAWSES-II period was extremely quiet, allowing the pure effects from the lower atmosphere to be observed in isolation without disturbances from above.

There are a number of critically important open questions toward the goal of achieving space weather predictability. Significant amounts of variability remain unexplained and/or are very difficult to detect and interpret, for example, the imprints of convectively forced gravity waves in the thermosphere. Short-term tidal variability in the neutrals and plasma is largely unknown because of the lack of suitable satellite data since local time coverage by a single satellite is limited. It is not clear what is more important: wave propagation into the thermosphere and then into to the ionosphere; or electrodynamic coupling between the E- and F-region and then into the thermosphere. The dynamics community still knows little about interactions between the various types of waves (tides, PWs, and GWs) and the mean flow. A major observational and technological challenge is the lack of global wind observations, day and night, throughout the 120 to 400 km region.

As overviewed below, new space-borne and ground-based assets, along with progress in geospace modeling, will help to close this knowledge gap that is also targeted through elements of SCOSTEP's new VarSITI program.

Program Outreach

As a dedicated effort of the international scientific community, and in accordance with SCOSTEP's mission not only to run scientific programs but also to promote solar-terrestrial physics and dissemination of the derived knowledge for the benefit of society, Task Group 4 published a quarterly newsletter. Each newsletter was about eight pages long with updates on recent campaign activities, short news, and a portrait of a young scientist. All articles were written to be understandable for non-specialist and published on the task group wiki (www.cawses.org/wiki/index.php/Task_4) and through an extensive mailing list. A total of 13 issues have been published with 64 articles from authors from 20 different countries, including 10 articles featuring young scientists. TG4 also organized and/or supported 12 dedicated workshops and special sessions and held annual business meetings during major conferences.

New and Future Satellite Missions

Space agencies and national science foundations in the U.S., Europe, and Asia by now recognize the importance of lower atmosphere driving of the ionosphere-thermosphere system not only as a domain of compelling scientific inquiry but also as highly relevant for understanding and predicting space weather; a task highly relevant for technological societies.

In the United States, the Decadal Survey on Solar and Space Physics 2013 to 2022 conducted by the National Academy of Sciences (National Research Council [2013]) recognized the realization that weather systems in the troposphere impact space weather through tides and gravity waves as one of the significant discoveries in the past decade, making 'the comprehensive understanding of the variability in space weather driven by lower-atmosphere weather on Earth' its second highest priority and recommended a dedicated satellite constellation (DYNAMIC) to be launched around the year 2020. While NASA budget constraints may delay the implementation of DYNAMIC; two new satellite missions dedicated to ionosphere-thermosphere physics, ICON and GOLD, will be launched in 2017. Both ICON and GOLD can be seen as pathfinder missions for DYNAMIC. ICON will collect data from a low Earth orbit to compare the impacts of direct solar driving and lower atmosphere driving on variability in the ionosphere/thermosphere system, and GOLD will image the thermosphere from a geostationary orbit, allowing for an unprecedented global view of short-term variability.

A very exciting new mission, launched on 22 November 2013, is the European Space Agency (ESA) Swarm mission. Swarm is a unique three satellite constellation with two satellites orbiting at 480 km and one at 510 km to measure Earth's magnetic field, the electric field in the ionosphere, and thermospheric density and winds. See the special issue in Earth, Planets and Space on Swarm Science Data Processing and Products - the Swarm Satellite Constellation Application and Research Facility ([2013]) for details. Each of the identical Swarm satellites has a vector and scalar magnetometer, a Langmuir probe, an accelerometer, and GPS instrumentation on board, similar to the tremendously successful CHAMP satellite. During its anticipated 10-year mission, the Swarm data will allow the ionosphere/thermosphere community to obtain an unprecedented view of small-scale structures in the thermosphere and the ionospheric plasma.

Due to the success of COSMIC (a particularly important dataset for the ionosphere/thermosphere community because of its global electron density profiles from radio occultation), the United States and Taiwan will launch COSMIC-2 between 2015 and 2018. COSMIC-2 consists of six satellites in a low-inclination orbit (launch in 2015) and another six satellites in high inclination orbits (launch in 2018).

Along with emerging cubesat and nanosat capabilities, these new space-borne assets will form the backbone of studying the signatures and impact of lower atmosphere variability versus magnetosphere/solar-driven variability in the next decade. Their synergistic use will most likely allow the community to make significant progress toward understanding geospace as a system and also toward predictability of space weather, a significant need for a technological society. A particular challenge, however, will remain: global thermospheric winds throughout the whole thermosphere during day- and nighttime; dataset airglow interferometers cannot provide. The implementation of new emerging technologies, such as the Doppler wind and temperature sounder (DWTS; Gordley and Marshall [2011]; Lieberman et al. [2012]), will be needed.

New Ground-based Assets

Ground-based facilities are powerful tools to measure vertical coupling from the atmosphere to geospace. They are basically radio techniques (various types of radars, ionosondes, satellite beacon receivers including Global Navigation Satellite System (GNSS) receivers, natural/artificial wave receivers, and magnetometers) and optical techniques (imagers and interferometers for airglow and aurora; and lidars). Figure 22 summarizes the altitude ranges and measurement parameters of various ground-based techniques. Because a single instrument covers only a certain range of altitudes and can measure a limited set of physical parameters, it is essential to combine several types of instruments at the same place to understand the connection from the troposphere to geospace. Multi-point networks of these various instruments are becoming available, providing global information of these parameters, e.g., Yumoto ([2011]) - MAGDAS and Yizengaw and Moldwin ([2009])

- AMBER for magnetometers, Makela et al. ([2009]) - RENOIR, Shiokawa et al. ([2009]) - OMTIs, Makela et al. ([2012]) - NATION, and Meriwether et al. ([2014]) - Peruvian FPIs for optical instruments, and Cohen et al. ([2010]) - AWESOME for LF/VLF receivers. It is also important to combine these ground-based observations with satellite measurements to separate temporal and spatial variations.

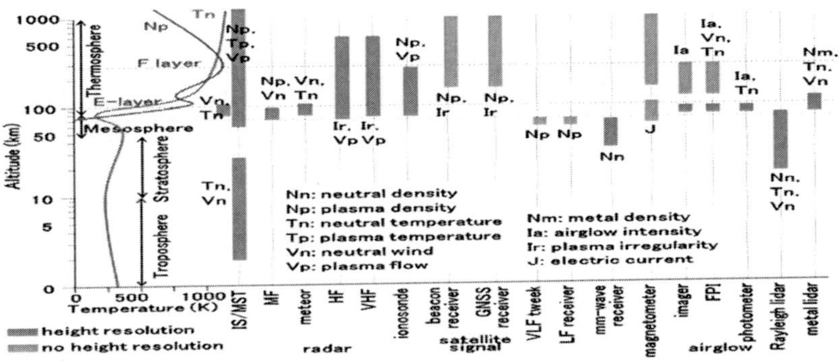

Figure 22: Various ground-based instruments to observe the upper atmosphere, their altitude ranges, and parameters of measurements.

One of the great progress of ground-based observation during CAWSES-II was that the multi-point GNSS receiver network is becoming a powerful tool to provide two-dimensional images of the ionospheric total electron content in the spatial resolution of less than approximately 100 km, as shown in Figure 2 as a striking example of tsunami- and tornado-induced gravity wave penetration to the ionosphere (Tsugawa et al. [2011]; Nishioka et al. [2013]). Such high-resolution TEC maps become available not only over Japan and US, but also in other places, e.g., over South America (Takahashi et al. [2014]) and Europe (Otsuka et al. [2013]). The number of GNSS receivers will grow further even after the CAWSES-II era and will become an important facility to monitor ionospheric disturbances.

The widest altitude coverage from the troposphere to the ionosphere can be achieved by powerful facilities of incoherent scatter (IS)/mesosphere-stratosphere-troposphere (MST) radars. Currently, major facilities of IS/MST radars are available at Resolute Bay (RISR), Poker Flat (PFISR), Sondrestrom, Milstone Hill, Arecibo, and Jicamarca in the American longitudinal sector; Irkutsk, Shigaraki

(MU), Wuhan, Gadanki, and Kototabang (EAR) in the Asian sector; and Svalbard (ESR), Scandinavia (EISCAT), Andøya (MAARSY), Kharkov, and Syowa (PANSY) in the European sector. The EISCAT radar will be renewed into the EISCAT_3D radar (Wannberg et al. [2010]) to obtain three-dimensional information of plasma-neutral interaction in the auroral ionosphere/thermosphere. This renewal will also stimulate measurements of tropospheric and stratospheric parameters using its powerful facilities. The Equatorial Atmosphere Radar (EAR) is planned to be renewed to the EMU (Equatorial Middle and Upper atmosphere) radar with more than five times larger transmission power to perform IS observations of the ionosphere.

VarSITI - SCOSTEP's 2014 to 2018 Program

The new 2014 to 2018 SCOSTEP program Variability of the Sun and Its Terrestrial Impact (VarSITI) will focus on the unusual low solar activity during solar cycle 24 and its consequences on Earth. One of the four VarSITI projects is called Role of the Sun and the Middle Atmosphere/Thermosphere/Ionosphere in Climate (ROSMIC) and is directed toward identifying the effects of external forcing on and predicting the effects of internal changes to the atmosphere/ionosphere. One goal of ROSMIC is to understand the impact of the Sun on the terrestrial middle atmosphere/lower thermosphere/ionosphere relative to anthropogenic forcing including coupling within the atmosphere-ionosphere system. As such, ROSMIC continues the science investigations related to understanding the geospace response to lower atmosphere variability that were carried out during CAWSES-II, using similar approaches such as observing campaigns, workshops, etc.

AUTHORS' CONTRIBUTIONS

JO, KS, and SG have been co-leaders of Task Group 4 (TG4) 'What is the geospace response to variable inputs from the lower atmosphere?' of SCOSTEP's CAWSES-II program. WW is the leader of the ROSMIC sub-program of SCOSTEP's VarSITI program and also led project 1. JM and HT organized the LONET campaign and led project 2. HF and MK led project 4. MK organized various EISCAT and PFISR campaigns. All authors read and approved the final manuscript.

ACKNOWLEDGEMENTS

JO, KS, and SG are grateful for the enthusiasm and contributions of all 250 members of TG4, especially the project leaders W. Ward, M. A. Abdu, J. Chau, J. Makela, H. Takahashi, D. Pancheva, M. Yamamoto, H. Fujiwara, and M. Kosch. Our special acknowledgement goes to Dr. Michi Nishioka who edited the quarterly newsletter. JM and HT thank all of the LONET and SpreadFEx-2 campaign participants. JM and HT acknowledge the COSMIC data center team who provided the TEC data. The Pameungpeuk MF radar data were provided through IUGONET under agreement of RISH, Kyoto University. JO was supported by NSF award 1139048 and NASA grants NNX11AJ13G and NNH12CF66C. KS is supported by the JSPS Grants-in-Aid for Scientific Research (20244080, 23403009, and 25247080), JSPS Core-to-Core Program, B. Asia-Africa Science Platforms, and the STEL Cooperative and IUGONET Projects from MEXT, Japan.

REFERENCES

1. Anderson C, Davies T, Conde M, Dyson P, Kosch MJ (2011) Spatial sampling of the thermospheric vertical wind field at auroral latitudes. J Geophys Res 116: doi:10.1029/2011JA016485
2. Anderson C, Kosch MJ, Nicolls MJ, Conde M (2013) Ion-neutral coupling in the earth's thermosphere, estimate from concurrent radar and optics observations over Alaska. J Atmos Sol Terr Phys 105–106:313-324 doi:10.1016/j.jastp.2013.04.005
3. Anthes RA, Ector D, Hunt DC, Kuo YH, Rocken C, Schreiner WS, et al. (2008) The COSMIC/FORMOSAT-3 mission: early results. Bull Amer Meteor Soc 89:313-333doi:10.1175/BAMS-89-3-313
4. Aruliah AL, Griffin EM, McWhiter I, Aylward AD, Ford EAK, Charalambous A, et al. (2004) First tristatic studies of meso-scale ion-neutral dynamics and energetic in the high-latitude upper atmosphere using co-located FPIs and EISCAT radar. Geophys Res Lett 31: doi:10.1029/2003GL018469
5. Aruliah AL, Griffin EM, Aylward AD, Ford EAK, Kosch MJ, Davis CJ, et al. (2005) First direct evidence of meso-scale variability on ion-neutral dynamics using co-located tristatic FPIs and EISCAT radar in northern Scandinavia. Ann Geophys 23:147-162

6. Booker HG, Wells HW (1938) Scattering of radio waves by the F-region of the ionosphere. J Geophys Res 43:249-256

7. Burch JL (2000) IMAGE mission overview. Space Science Reviews 91(1–14):2000

8. Chang LC, Ward WE, Palo SE, Du J, Wang DY, Liu HL, et al. (2012) Comparison of diurnal tide in models and ground-based observations during the 2005 equinox CAWSES tidal campaign. J Atmos Sol Terr Phys 78–79:19-30 doi:10.1016/j.jastp.2010.12.0104

9. Chang LC, Lin CH, Yue J, Liu JY, Lin JT (2013) Stationary planetary wave and nonmigrating tidal signatures in ionospheric wave 3 and wave 4 variations in 2007-2011 FORMOSAT-3/COSMIC observations. J Geophys Res Space Phys 118(10):6651-6665 doi:10.1002/jgra.50583

10. Chapagain NP, Makela JJ, Meriwether JW, Fisher DJ, Buriti RA, Medeiros AF (2012) Comparison of nighttime zonal neutral winds and equatorial plasma bubble drift velocities over Brazil. J Geophys Res 117: doi:10.1029/2012JA017620

11. Chau JL, Fejer BG, Goncharenko LP (2009) Quiet variability of equatorial E x B drifts during a sudden stratospheric warming event. Geophys Res Lett 36: doi:10.1029/2008GL036785

12. Chau JL, Aponte NA, Cabassa E, Sulzer MP, Goncharenko LP, Gonzalez SA (2010) Quiet time ionospheric variability over Arecibo during sudden stratospheric warming events. J Geophys Res 115:A00G06 doi:10.1029/2010JA015378

13. Chau JL, Goncharenko LP, Fejer BG, Liu HL (2012) Equatorial and low latitude ionospheric effects during sudden stratospheric warming events. Space Sci Rev 168:385-417doi:10.1007/s11214-011-9797-5

14. Chen G, Xu J, Wang W, Lei J, Burns AG (2012) A comparison of the effects of CIR- and CME-induced geomagnetic activity on thermospheric densities and spacecraft orbits: case studies. J Geophys Res 117:doi:10.1029/2012JA017782

15. Chu X, Yu Z, Gardner CS, Chen C, Fong W (2011) Lidar observations of neutral Fe layers and fast gravity waves in the thermosphere (110-155 km) at McMurdo (77.8°S, 166.7°E), Antarctica. Geophys Res Lett 38(23): doi:10.1029/2011GL050016

16. Cohen MB, Inan US, Paschal EW (2010) Sensitive broadband ELF/VLF radio reception with the AWESOME instrument. IEEE Trans On Geosci Remote Sens 48(1):3-17 doi:10.1109/TGRS.2009.2028334
17. de Larquier S, Pasko VP, Stenbaek-Nielsen HC, Wilson CR, Olson JV (2010) Finite-difference time-domain modeling of infrasound from pulsating auroras and comparison with recent observations. Geophys Res Lett 37: doi:10.1029/2009GL042124
18. Deng Y, Fuller-Rowell TJ, Akmaev RA, Ridley AJ (2011) Impact of the altitudinal Joule heating distribution on the thermosphere. J Geophys Res 116: doi:10.1029/2010JA016019
19. Dungey JW (1956) Convective diffusion in the equatorial F-region. J Atmos Sol Terr Phys 9:304-310
20. Eastes RW, McClintock WE, Codrescu MV, Aksnes A, Anderson D, Andersson L, et al. (2008) Global-scale Observations of the Limb and Disk (GOLD): new observing capabilities for the ionosphere-thermosphere. In: Kintner PM, Coster AJ, Fuller-Rowell T, Mannucci J, Mendillo M, Heelis R (eds) Midlatitude ionospheric dynamics and disturbances, American Geophysical Union, Washington DC. pp 319-326
21. England SL (2012) Review of the effects of non-migrating atmospheric tides on Earth's low-latitude ionosphere. Space Sci Rev 168(1–4):211-236 doi:10.1007/s11214-011-9842-4
22. England SL, Immel TJ, Huba JD, Hagan ME, Maute A, DeMajistre R (2010) Modeling of multiple effects of atmospheric tides on the ionosphere: an examination of possible coupling mechanisms responsible for the longitudinal structure of the equatorial ionosphere. J Geophys Res 115: doi:10.1029/2009JA014894
23. Fejer JA (1964) Atmospheric tides and associated magnetic effects. Rev Geophys 2(2):275-309
24. Fejer BG, Olson ME, Chau JL, Stolle C, Lühr H, Goncharenko LP, et al. (2010) Lunar-dependent equatorial ionospheric electrodynamic effects during sudden stratospheric warmings. J Geophys Res 115:A00G03 doi:10.1029/2010JA015273
25. Forbes JM (1995) Tidal and Planetary Waves. In: Johnson RM, Killeen TL (eds) The upper mesosphere and lower thermosphere: a review of experimental and theory. geophysical monograph series, American Geophysical Union, Washington DC. pp 67-87

26. Forbes JM, Russell J, Miyahara S, Zhang X, Palo S, Mlynczak M, et al. (2006) Troposphere-thermosphere tidal coupling as measured by the SABER instrument on TIMED during July-September 2002. J Geophys Res 111:A10S06 doi:10.1029/2005JA011492
27. Forbes JM, Bruinsma SL, Zhang X, Oberheide J (2009) Surface-exosphere coupling due to thermal tides. Geophys Res Lett 36: doi:10.1029/2009GL038748
28. Forbes JM, Zhang X, Bruinsma S, Oberheide J (2011) Sun-synchronous thermal tides in exosphere temperature from CHAMP and GRACE accelerometer measurements. J Geophys Res 116: doi:10.1029/2011JA016855
29. Forbes JM, Zhang X, Bruinsma SL (2014) New perspectives on thermospheric tides: 2. penetration to the upper thermosphere. Earth Planets Space 66:122 doi:10.1186/1880-5981-66-122
30. Fritts DC, Lund TS (2011) Gravity wave influences in the thermosphere and ionosphere: observations and recent modeling. In: Abdu MA, Pancheva D (eds) Aeronomy of the Earth's atmosphere and ionosphere. IAGA Special Sopron Book Series, Springer Science+Business Media, . pp 109-130 doi:10.1007/978-94-007-0326-1_8
31. Fritts DC, Vadas SL, Riggin DM, Abdu MA, Batista IS, Takahashi H, et al. (2008) Gravity wave and tidal influences on equatorial spread F based on observations during the spread F experiment (SpreadFEx). Ann Geophys 26:3235-3252 doi:10.5194/angeo-26-3235-2008
32. Fujitaka K, Tohmatsu T (1973) A tidal theory of the ionospheric intermediate layer. J Atmos Terr Phys 33:425-438
33. Fujiwara H, Nozawa S, Maeda S, Ogawa Y, Miyoshi Y, Jin H, et al. (2012) Polar cap thermosphere and ionosphere during the solar minimum period: EISCAT Svalbard radar observations and GCM simulations. Earth Planets Space 64:459-465 doi:10.5047/eps.2011.05.023
34. Fukushima D, Shiokawa K, Otsuka Y, Ogawa T (2012) Observation of equatorial nighttime medium-scale traveling ionospheric disturbances in 630-nm airglow images over 7 years. J Geophys Res 117: doi:10.1029/2012JA017758
35. Fuller-Rowell T, Wu F, Akmaev R, Fang TW, Araujo-Pradere E (2010) A whole atmosphere model simulation of the

impact of a sudden stratospheric warming on thermosphere dynamics and electrodynamics. J Geophys Res 115:A00G08 doi:10.1029/2010JA015524

36. Fuller-Rowell T, Wang H, Akmaev R, Wu F, Fang TW, Iredell M, et al. (2011) Forecasting the dynamic and electrodynamic response to the January 2009 sudden stratospheric warming. Geophys Res Lett 38: doi:10.1029/2011GL047732
37. Fytterer T, Arras C, Hoffmann P, Jacobi C (2014) Global distribution of the migrating terdiurnal tide seen in sporadic E occurrence frequencies obtained from GPS radio occultations. Earth Planets Space 66:79 doi:10.1186/1880-5981-66-79
38. Gan Q, Du J, Ward WE, Beagley SR, Fomichev VI, Zhang S (2014) Climatology of the diurnal tides from eCMAM30 (1979 to 2010) and its comparisons with SABER. Earth Planets Space 66:103 doi:10.1186/1880-5981-66-103
39. Gardner LC, Schunk RW (2011) Large-scale gravity wave characteristics simulated with a high-resolution global thermosphere-ionosphere model. J Geophys Res 116: doi:10.1029/2010JA015629
40. Gardner L, Sojka JJ, Schunk RW, Heelis R (2012) Changes in thermospheric temperature induced by high-speed solar wind streams. J Geophys Res 117: doi:10.1029/2012JA017892
41. Gavrilov NM, Kshetvetskii SP (2014) Three-dimensional numerical simulation of nonlinear acoustic-gravity wave propagation from the troposphere to the thermosphere. Earth Planets Space 66:88 doi:10.1186/1880-5981-66-8
42. Goncharenko L, Zhang SR (2008) Ionospheric signatures of sudden stratospheric warming: Ion temperature at middle latitude. Geophys Res Lett 35: doi:10.1029/2008GL035684
43. Goncharenko LP, Chau JL, Liu HL, Coster AJ (2010) Unexpected connections between the stratosphere and ionosphere. Geophys Res Lett 37: doi:10.1029/2010GL043125
44. Goncharenko LP, Hsu VW, Brum CGM, Zhang SR, Fentzke JT (2013) Wave signatures in the midlatitude ionosphere during a sudden stratospheric warming of January 2010. J Geophys Res 118:472-487 doi:10.1029/2012JA018251

45. Gordley LL, Marshall BT (2011) Doppler wind and temperature sounder (DWTS): a new approach using gas filter radiometry. J Appl Rem Sens 5:053570 doi:10.1117/1.3666048

46. Haagmans R, Bock R, Rider H (2013) Swarm: ESA's magnetic field mission. ESA communications, BR-302, September 2013.

47. Hagan ME, Forbes JM (2002) Migrating and nonmigrating diurnal tides in the middle and upper atmosphere excited by tropospheric latent heat release. J Geophys Res 107(D24):4754 doi:10.1029/2001JD001236

48. Hagan ME, Forbes JM (2003) Migrating and nonmigrating semidiurnal tides in the upper atmosphere excited by tropospheric latent heat release. J Geophys Res 108(A2):1062 doi:10.1029/2002JA009466

49. Hagan ME, Roble RG (2001) Modeling the diurnal tidal variability with the National Center for Atmospheric Research thermosphere-ionosphere-mesosphere-electrodynamics general circulation model. J Geophys Res 106(A11):24,869-24,882

50. Hagan ME, Maute AI, Roble RG, Richmond AD, Immel TJ, England SL (2007) Connections between deep tropical clouds and the Earth's ionosphere. Geophys Res Lett 34: doi:10.1029/2007GL030142

51. Hagan ME, Maute AI, Roble RG (2009) Tropospheric tidal effects on the middle and upper atmosphere. J Geophys Res 114: doi:10.1029/2008JA013637

52. Haldoupis C (2011) Midlatitude sporadic *E*. A typical paradigm of atmosphere-ionosphere coupling. Space Sci Rev 168:441-461 doi:10.1007/s11214-011-9786-8

53. Häusler K, Oberheide J, Lühr H, Koppmann R (2013) The geospace response to nonmigrating tides. In: Lübken FJ (ed) Climate and Weather of the Sun-Earth System (CAWSES), Springer Atmospheric Sciences Series, pp. 481-504

54. Hines CO (1960) Internal gravity waves at ionospheric heights. Can J Phys 38(11):1441-1481

55. Immel TJ, Sagawa E, England SL, Henderson SB, Hagan ME, Mende SB, et al. (2006) Control of equatorial ionospheric morphology by atmospheric tides. Geophys Res Lett 33: doi:10.1029/2006GL026161

56. International CAWSES-II Symposium (2014). Special Issue in Earth Planets Space 66. In: Yamamoto M, Shiokawa K, Nakamura T, Gopalswamy N (eds)
57. Jin H, Miyoshi Y, Fujiwara H, Shinagawa H, Terada K, Terada N, et al. (2011) Vertical connection from the tropospheric activities to the ionospheric longitudinal structure simulated by a new Earth's whole atmosphere-ionosphere coupled model. J Geophys Res 116: doi:10.1029/2010JA015925
58. Jin H, Miyoshi Y, Pancheva D, Mukhtarov P, Fujiwara H, Shinagawa H (2012) Response of migrating tides to the stratospheric sudden warming in 2009 and their effects on the ionosphere studied by a whole atmosphere-ionosphere model GAIA with COSMIC and TIMED/SABER observations. J Geophys Res 117(A10): doi:10.1029/2012JA017650
59. Jones M Jr, Forbes JM, Hagan ME, Maute A (2013) Non-migrating tides in the ionosphere-thermosphere: in situ versus tropospheric sources. J Geophys Res Space Phys 118(5):2438-2451 doi:10.1002/jgra.50257
60. Kishore Kumar G, Singer W, Oberheide J, Grieger N, Batista PP, Riggin DM, et al. (2013) Diurnal tides at low latitudes: radar, satellite, and model results. J Atmos Sol Terr Phys 118:96-105 doi:10.1016/j.jastp.2013.07.005
61. Kosch MJ, Anderson C, Makarevich RA, Carter BA, Fiori RAD, Conde M, et al. (2010) First E region observations of mesoscale neutral wind interaction with auroral arcs. J Geophys Res 115: doi:10.1029/2009JA014697
62. Kosch M, Yiu I, Anderson C, Tsuda T, Ogawa Y, Nozawa S, et al. (2011) Meso-scale observations of Joule heating near an auroral arc and ion-neutral collision frequency in the polar cap E-region. J Geophys Res 116: doi:10.1029/2010JA016015
63. Kubota M, Conde M, Ishii M, Murayama Y, Jin H (2011) Characteristics of nighttime MSTID/AGW observed over Alaska. J Geophys Res 116: doi:10.1029/2010JA016212
64. Kurihara J, Oyama S, Nozawa S, Tsuda TT, Fujii R, Ogawa Y, et al. (2009) Temperature enhancements and vertical winds in the lower thermosphere associated with auroral heating during the DELTA campaign. J Geophys Res 114: doi:10.1029/2009JA014392

65. Laskar FI, Pallamraju D, Veenadhari B (2014) Vertical coupling of atmospheres: dependence on strength of sudden stratospheric warming and solar activity. Earth Planets Space 66:94 doi:10.1186/1880-5981-66-94

66. Lastovicka J (2006) Forcing of the ionosphere by waves from below. J Atmos Sol Terr Phys 68:479-497 doi:10.1016/j.jastp.2005.01.018

67. Lei J, Thayer JP, Wang W, McPherron RL (2011) Impact of CIR storms on thermosphere density variability during the solar minimum of 2008. Sol Phys 274:427-437 doi:10.1007/s11207-010-9563-y

68. Lieberman RS, Oberheide J, Gordley LL, Marshall BT (2012) Recovery of planetary-scale waves in stratospheric, mesospheric, and lower thermospheric winds and temperature from the Doppler wind and temperature sounder. J Appl Remote Sens 6(1):063570 doi:101117/1.JRS.6.063570

69. Lieberman RS, Oberheide J, Talaat ER (2013) Nonmigrating diurnal tides observed in global thermospheric winds. J Geophys Res Space Phys 118:7384-7397 doi:10.1002/2013JA018975

70. Lieberman RS, Riggin DM, Siskind DE (2013) Stationary waves in the wintertime mesosphere: evidence for gravity wave filtering by stratospheric planetary waves. J Geophys Res Atmos 118:3139-3149 doi:10.1002/jgrd.50319

71. Lin CH, Wang W, Hagan ME, Hsiao CC, Immel TJ, Hsu ML, et al. (2007) Plausible effect of atmospheric tides on the equatorial ionosphere observed by the FORMOSAT-3/COSMIC: three-dimensional electron density structures. Geophys Res Lett 34: doi:10 1029/2007GL029265

72. Liu HL, Roble RG (2002) A study of a self-generated stratospheric sudden warming and its mesospheric/lower thermospheric impacts using the coupled TIMEGCM/CCM3. J Geophys Res 107(D23):1-18 doi:10.1029/2001JD001533

73. Liu H, Yamamoto M (2011) Weakening of the mid-latitude summer nighttime anomaly during geomagnetic storms. Earth Planets Space 63:371-375 doi:10.5047/eps.2010.11.012

74. Liu H, Yamamoto M, Lühr H (2009) Wave-4 pattern of the equatorial mass density anomaly: a thermospheric signature of tropical deep convection. Geophys Res Lett 36: doi:10.1029/2009GL039865

75. Liu G, Immel TJ, England SL, Kumar KK, Ramkumar G (2010) Temporal modulations of the longitudinal structure in F2 peak height in the equatorial ionosphere as observed by COSMIC. J Geophys Res 115: doi:10.1029/2009JA014829

76. Liu G, Immel TJ, England SL, Kumar KK, Ramkumar G (2010) Temporal modulation of the four-peaked longitudinal structure of the equatorial ionosphere by the 2 day planetary wave. J Geophys Res 115: doi:10.1029/2010JA016071

77. Liu HL, Foster BT, Hagan ME, McInerney JM, Maute A, Qian L, et al. (2010) Thermosphere extension of the whole atmosphere community climate model. J Geophys Res 115: doi:10.1029/2010JA015586

78. Liu HL, Wang W, Richmond AD, Roble RG (2010) Ionospheric variability due to planetary waves and tides for solar minimum conditions. J Geophys Res 115:A00G01 doi:10.1029/2009JA015188

79. Liu H, Jin H, Miyoshi Y, Fujiwara H, Shinagawa H (2013) Upper atmosphere response to stratosphere sudden warming: local time and height dependence simulated by GAIA model. Geophys Res Lett 40:635-640 doi:10.1002/grl.50146

80. Liu HL, Yudin VA, Roble RG (2013) Day-to-day ionospheric variability due to lower atmosphere perturbations. Geophys Res Lett 40:665-670 doi:10.1002/grl.50125

81. Lühr H, Rother M, Kohler W, Ritter P, Grunwaldt L (2004) Thermospheric up-welling in the cusp region: evidence from CHAMP observations. Geophys Res Lett 31(6): doi:10.1029/2003GL019314

82. Lühr H, Park J, Ritter P, Liu H (2012) In-situ CHAMP observations of the ionosphere-thermosphere coupling. Space Sci Rev 168(1–4):237-260 doi:10.1007/s11214-011-9798-4

83. Makela JJ, Meriwether JW, Lima JP, Miller ES, Armstrong SJ (2009) The remote equatorial nighttime observatory of ionospheric regions project and the international heliospherical year. Earth Moon Planets 104(1–4):211-226 doi:10.1007/s11038-008-9289-0

84. Makela JJ, Vadas SL, Muryanto R, Duly T, Crowley G (2010) Periodic spacing between consecutive equatorial plasma bubbles. Geophys Res Lett 37: doi:10.1029/2010GL043968

85. Makela JJ, Lognonné P, Hébert H, Gehrels T, Rolland L, Allgeyer S, et al. (2011) Imaging and modeling of the ionospheric airglow response over Hawaii to the tsunami generated by the Tohoku earthquake of 11 March 2011. Geophys Res Lett 38:L00G02 doi:10.1029/2011GL047860

86. Makela JJ, Meriwether JW, Ridley AJ, Ciocca M, Castelaz MW (2012) Large-scale measurements of thermospheric dynamics with a multisite Fabry-Pérot interferometer network: overview of plans and results from midlatitude measurements. Int J Geophys Article ID :Article ID 872140 doi:10.1155/2012/872140

87. Makela JJ, Fisher DJ, Meriwether JW, Buriti RA, Medeiros AF (2013) Near-continual ground-based nighttime observations of thermospheric neutral winds and temperatures over equatorial Brazil from 2009 to 2012. J Atmos Sol Terr Phys 103:94-102 doi:10.1016/j.jastp.2012.11.019

88. Matsumura M, Saito A, Ieymori T, Shinagawa H, Tsugawa T, Otsuka Y, et al. (2011) Numerical simulations of atmospheric waves excited 2011 off the Pacific coast of Tohoku earthquake. Earth Planets Space 63(7):885-889 doi:10.5047/eps.2011.07.015

89. Maute A, Richmond AD, Roble RG (2012) Sources of low-latitude ionospheric ExB drifts and their variability. J Geophys Res 117: doi:10.1029/2011JA017502

90. Meriwether JW, Makela JJ, Fisher DJ, Buriti RA, Medeiros AF, Akmaev RA, et al. (2013) Comparisons of thermospheric wind and temperature measurements in equatorial brazil to whole atmosphere model predictions. J Atmos Sol Terr Phys 103:103-112 doi:10.1016/j.jastp.2013.04.002

91. Meriwether JW, Makela JJ, Ridley A, Hampton DL, Gerrard A, Ciocca M, et al. (2014) Observations from a network of Fabry-Pérot interferometers and the HOPE-A probe. Paper presented at AGU Chapman Conference on Magnetosphere-Ionosphere Coupling in the Solar System, Yosemite National Park, CA.

92. Mukhtarov P, Pancheva D, Andonov B (2010) Climatology of the stationary planetary waves seen in the SABER/TIMED temperatures (2002-2007). J Geophys Res 115: doi:10.1029/2009JA015156

93. Narayanan VL, Sau S, Gurubaran S, Shiokawa K, Balan N, Emperumal K, et al. (2014) A statistical study of satellite traces and subsequent evolution of equatorial spread F based on ionosonde

observations over dip equatorial site Tirunelveli, India. Earth Planets Space 66:160

94. (2013) Solar and space physics: a science for a technological society. The National Academies Press, Washington DC.
95. Nguyen V, Palo SE (2013) Technique to produce daily estimates of the migrating diurnal tide using TIMED/SABER and EOS Aura/MLS. J Atmos Sol Terr Phys 105–106:39-53 doi:10.1016/j.jastp.2013.07.008
96. Nicolls MJ, Heinselman CJ (2007) Three-dimensional measurements of traveling ionospheric disturbances with the Poker Flat Incoherent Scatter Radar. Geophys Res Lett 34: doi:10.1029/2007GL031506
97. Nishioka M, Tsugawa T, Kubota M, Ishii M (2013) Concentric waves and short-period oscillations observed in the ionosphere after the 2013 Moore EF5tornado. Geophys Res Lett 40:5581-5586 doi:10.1002/2013GL057963
98. Oberheide J, Forbes JM (2008) Thermospheric nitric oxide variability induced by nonmigrating tides. Geophys Res Lett 35: doi:10.1029/2008GL034825
99. Oberheide J, Hagan ME, Roble RG, Offermann D (2002) Sources of nonmigrating tides in the tropical middle atmosphere. J Geophys Res 107(D21):4567 doi:10.1029/2002JD002220
100. Oberheide J, Liu HL, Gusev O, Offermann D (2006) Mesospheric surf zone and temperature inversion layers in early November 1994. J Atmos Sol Terr Phys 68(15):1752-1763 doi:10.1016/j.jastp.2005.11.013
101. Oberheide J, Wu Q, Killeen TL, Hagan ME, Roble RG (2006) Diurnal nonmigrating tides from TIMED Doppler interferometer wind data: monthly climatologies and seasonal variations. J Geophys Res 111:A10S03 doi:10.1029/2005JA011491
102. Oberheide J, Forbes JM, Häusler K, Wu Q, Bruinsma SL (2009) Tropospheric tides from 80-400 km: propagation, inter-annual variability and solar cycle effects. J Geophys Res 114:D00I05 doi:10.1029/2009JD012388
103. Oberheide J, Forbes JM, Zhang X, Bruinsma SL (2011) Climatology of upward propagating diurnal and semidiurnal tides in the thermosphere. J Geophys Res 116: doi:10.1029/2011JA016784

104. Oberheide J, Forbes JM, Zhang X, Bruinsma SL (2011) Wave-driven variability in the ionosphere-thermosphere-mesosphere system from TIMED observations: what contributes to the 'wave-4'? J Geophys Res 116: doi:10.1029/2010JA015911

105. Oberheide J, Mlynczak MG, Mosso CN, Schroeder BM, Funke B, Maute A (2013) Impact of tropospheric tides on the nitric oxide 5.3 μm infrared cooling of the low-latitude thermosphere during solar minimum conditions. J Geophys Res 118:7283-7293 doi:10.1002/2013JA019278

106. Otsuka Y, Shiokawa K, Ogawa T (2012) Disappearance of equatorial plasma bubble after interaction with mid-latitude medium-scale traveling ionospheric disturbance. Geophys Res Lett 39: doi:10.1029/2012GL052286

107. Otsuka Y, Suzuki K, Nakagawa S, Nishioka M, Shiokawa K, Tsugawa T (2013) GPS observations of medium-scale traveling ionospheric disturbances over Europe. Ann Geophys 31:163-172 doi:10.5194/angeo-31-163-2013

108. Oyama S, Watkins BJ (2012) Generation of atmospheric gravity waves in the polar thermosphere in response to auroral activity. Space Sci Rev 168(1–4):463-473 doi:10.1007/s11214-011-9847-z

109. Oyama S, Tsuda TT, Sakanoi T, Obuchi Y, Asamura K, Hirahara M, et al. (2009) Spatial evolution of frictional heating and the predicted thermospheric-wind effects in the vicinity of an auroral arc measured with the Sondrestrom incoherent-scatter radar and the Reimei satellite. J Geophys Res 114: doi:10.1029/2009JA014091

110. Oyama S, Shiokawa K, Kurihara J, Tsuda TT, Nozawa S, Ogawa Y, et al. (2010) Lower-thermospheric wind fluctuations measured with an FPI during pulsating aurora at Tromso, Norway. Ann Geophys 28:1847-1857 doi:10.5194/angeo-28-1847-2010

111. Pancheva D, Mukhtarov P (2011) Stratospheric warmings; the atmosphere-ionosphere coupling paradigm. J Atmos Sol Terr Phys 73:1697-1702 doi:10.1016/j.jastp.2011.03.006

112. Pancheva D, Mukhtarov P (2012) Planetary wave coupling of the atmosphere-ionosphere system during the Northern winter of 2008/2009. Adv Space Res 50:1189-1203 doi:10.1016/j.asr.2012.06.023

113. Pancheva D, Mukhtarov P, Andonov B, Forbes JM (2010) Global distribution and climatological features of the 5-6-day planetary waves seen in the SABER/TIMED temperatures (2002-2007). J Atmos Sol Terr Phys 72:26-37 doi:10.1016/j.jastp.2009.10.005

114. Park J, Lühr H, Kunze M, Fejer BG, Min KW (2012) Effect of sudden stratospheric warming on lunar tidal modulation of the equatorial electrojet. J Geophys Res 117: doi:10.1029/2011JA017351

115. Paulino I, Takahashi H, Medeiros AF, Wrasse CM, Buriti RA, Sobral JHA, et al. (2011) Mesospheric gravity waves and ionospheric plasma bubbles observed during the COPEX campaign. J Atmos Sol Terr Phys 73:1575-1580 doi:10.1016/j.jastp.2010.12.004

116. Pedatella NM, Forbes JM (2009) Interannual variability in the longitudinal structure of the low-latitude ionosphere due to the El Niño - Southern Oscillation. J Geophys Res 114: doi:10.1029/2009JA014494

117. Pedatella NM, Forbes JM (2009) Modulation of the equatorial F-region by the quasi-16-day planetary wave. Geophys Res Lett 36: doi:10.1029/2009GL037809

118. Pedatella NM, Forbes JM (2011) Electrodynamic response of the ionosphere to high-speed solar wind streams. J Geophys Res 116: doi:10.1029/2011JA017050

119. Pedatella NM, Liu HL (2012) Tidal variability in the mesosphere and lower thermosphere due to the El Niño - Southern Oscillation. Geophys Res Lett 39: doi:10.1029/2012GL053383

120. Pedatella NM, Liu HL, Sassi F, Lei J, Chau JL, Zhang X (2014) Ionosphere variability during the 2009 SSW: influence of the lunar semidiurnal tide and mechanisms producing electron density variability. J Geophys Res 119: (in press), doi:10.1002/2014JA019849

121. Perkins F (1973) Spread F and ionospheric currents. J Geophys Res 78:218-226

122. Qian L, Solomon SC, Mlynczak MG (2010) Model simulation of thermospheric response to recurrent geomagnetic forcing. J Geophys Res 115: doi:10.1029/2010JA015309

123. Reigber C, Lühr H, Grunwaldt L, Förste C, König R, Massmann H et al (2006) CHAMP Mission 5 Years in Orbit. In: Flury J, Reigber C, Rothacher M, Boedecker G, Schriber U (eds) Observation of

the Earth System from Space, Springer, pp 3-15, doi:10.1007/3-540-29522-4_1.

124. Ritter P, Lühr H, Doornbos E (2010) Substorm-related thermospheric density and wind disturbances derived from CHAMP observations. Ann Geophys 28:1207-1220 doi:10.5194/angeo-28-1207-2010

125. Sagawa E, Immel TJ, Frey HU, Mende SB (2005) Longitudinal structure of the equatorial anomaly in the nighttime ionosphere observed by IMAGE/FUV. J Geophys Res 110: doi:10.1029/2004JA010848

126. Shiokawa K, Otsuka Y, Ogawa T (2009) Propagation characteristics of nighttime mesospheric and thermospheric waves observed by optical mesosphere thermosphere imagers at middle and low latitudes. Earth Planets Space 61:479-491

127. Shiokawa K, Mori M, Otsuka Y, Oyama S, Nozawa S (2012) Motion of high-latitude nighttime medium-scale traveling ionospheric disturbances associated with auroral brightening. J Geophys Res 117: doi:10.1029/2012JA017928

128. Shiokawa K, Mori M, Otsuka Y, Oyama S, Nozawa S, Suzuki S, et al. (2013) Observation of nighttime medium-scale travelling ionospheric disturbances by two 630-nm airglow imagers near the auroral zone. J Atmos Sol Terr Phys 103:184-194 doi:10.1016/j.jastp.2013.03.024

129. Shiokawa K, Otsuka Y, Lynn KJ, Wilkinson P, Tsugawa T (2104) Airglow-imaging observation of plasma bubble disappearance at geomagnetically conjugate points. Earth Planets Space (in press)

130. Smith AK (2003) The origin of stationary waves in upper mesospheric winds. J Atmos Sci 54:2129-2145 doi:10.1175/1520-0469(2003)060<3033:TOOSPW>2.0.CO;2

131. Smith SM, Vadas SL, Baggaley WJ, Hernandez G, Baumgardner J (2013) Gravity wave coupling between the mesosphere and thermosphere over New Zealand. J Geophys Res Space Phys 118(5):2694-2707 doi:10.1002/jgra.50263

132. Stening RJ (2011) Lunar tide in the equatorial electrojet in relation to stratospheric warmings. J Geophys Res 116: doi:10.1029/2011JA017047

133. Stewart B (1882) Terrestrial magnetism. Encyclopedia Britannica.

134. (2013) . Special issue in Earth Planets Space 65(11):1185-1398.

135. Takahashi H, Taylor MJ, Pautet PD, Medeiros AF, Gobbi D, Wrasse CM, et al. (2009) Simultaneous observation of ionospheric plasma bubbles and mesospheric gravity waves during the SpreadFEx Campaign. Ann Geophys 27:1477-1487

136. Takahashi H, Costa S, Otsuka Y, Shiokawa K, Monico JFG, Paula E, et al. (2014) Diagnostics of equatorial and low latitude ionosphere by TEC mapping over Brazil. Adv Space Res 54(3):385-394 doi:10.1016/j.asr.2014.01.032

137. Taori A, Makela JJ, Taylor M (2010) Mesospheric wave signatures and equatorial plasma bubbles: a case study. J Geophys Res 115: doi:10.1029/2009JA015088

138. Taori A, Patra AK, Joshi LM (2011) Gravity wave seeding of equatorial plasma bubbles: an investigation with simultaneous F region, E region, and middle atmospheric measurements. J Geophys Res 116: doi:10.1029/2010JA016229

139. Tapley BD, Bettadpur S, Watkins M, Reigber C (2004) The gravity recovery and climate experiment: mission overview and early results. Geophys Res Lett 31: doi:10.1029/2004GL019920

140. Thampi SV, Yamamoto M, Tsunoda RT, Otsuka Y, Tsugawa T, Uemoto J, et al. (2009) First observations of large-scale wave structure and equatorial spread F using CERTO radio beacon on the C/NOFS satellite. Geophys Res Lett 36: doi:10.1029/2009GL039887

141. Thayer JP, Lei J, Forbes JM, Sutton EK, Nerem RS (2008) Thermospheric density oscillations due to periodic solar wind high-speed streams. J Geophys Res 113: doi:10.1029/2008JA013190

142. Tsuda TT, Nozawa S, Kawahara TD, Kawabata T, Saito N, Wada S, et al. (2013) Decrease in sodium density observed during auroral particle precipitation over Tromso, Norway. Geophys Res Lett 40:4486-4490 doi:10.1002/grl.50897

143. Tsugawa T, Saito A, Otsuka Y, Nishioka M, Maruyama T, Kato H, et al. (2011) Ionospheric disturbances detected by GPS total electron content observation after the 2011 off the Pacific coast of Tohoku earthquake. Earth Planets Space 63:875-879 doi:10.5047/eps.2011.06.035

144. Tsunoda RT (2005) On the enigma of day-to-day variability in equatorial spread F. Geophys Res Lett 32: doi:10.1029/2005GL022512

145. Tsunoda RT, Bubenik DM, Thampi SV, Yamamoto M (2010) On large-scale wave structure and equatorial spread F without a post-sunset rise of the F layer. Geophys Res Lett 37: doi:10.1029/2009GL042357

146. Tsunoda RT, Yamamoto M, Tsugawa T, Hoang TL, Tulasi Ram S, Thampi SV, et al. (2011) On seeding, large-scale wave structure, equatorial spread F, and scintillations over Vietnam. Geophys Res Lett 38: doi:10.1029/2011GL049173

147. Vadas SL, Crowley G (2010) Source of the traveling ionospheric disturbances observed by the ionospheric TIDDBIT sounder near Wallops Island on 30 October 2007. J Geophys Res 115: doi:10.1029/2009JA015053

148. Vadas SL, Liu HL (2009) The generation of large-scale gravity waves and neutral winds in the thermosphere from the dissipation of convectively-generated gravity waves. J Geophys Res 114: doi:10.1029/2009JA014108

149. Vadas SL, Liu HL (2013) Numerical modeling of the large-scale neutral and plasma responses to the body forces created by the dissipation of gravity waves from 6 h of deep convection in Brazil. J Geophys Res Space Phys 118:2593-2617 doi:10.1002/jgra.50249

150. Verbanac G, Vršnak B, Živkovi S, Hojsak T, Veronig AM, Temmer M (2011) Solar wind high-speed streams and related geomagnetic activity in the declining phase of solar cycle 23. Astro Astrophys 533: doi:10.1051/0004-6361/201116615

151. Vickers H, Kosch MJ, Sutton E, Ogawa Y, La Hoz C (2013) Thermospheric atomic oxygen density estimates using the EISCAT Svalbard radar. J Geophys Res 118:1319-1330 doi:10.1002/jgra.50169

152. Vickers H, Kosch MJ, Sutton E, Bjoland L, Ogawa Y, La Hoz C (2014) A solar cycle of upper thermosphere density observations from the EISCAT Svalbard radar. J Geophys Res Space Physics 119:6833-6845 doi:10.1002/2014JA019885

153. Wang H, Akmaev RA, Fang TW, Fuller-Rowell TJ, Wu F, Maruyama N, et al. (2014) First forecast of a sudden stratospheric warming with a coupled whole-atmosphere/ionosphere model IDEA. J Geophys Res Space Phys 119:2079-2089 doi:10.1002/2013JA019481

154. Wannberg UG, Andersson H, Behlke R, Belyey V, Bergqvist P, Borg J, et al. (2010) EISCAT_3D: a next-generation European radar system for upper-atmosphere and geospace research. Radio Sci Bull 332:75-88

155. Ward WE, Oberheide J, Goncharenko LP, Nakamura T, Hoffmann P, Singer W, et al. (2010) On the consistency of model, ground-based, and satellite observations of tidal signatures: Initial results from the CAWSES tidal campaigns. J Geophys Res 115: doi:10.1029/2009JD012593

156. Warner K, Oberheide J (2014) Nonmigrating tidal heating and MLT tidal wind variability due to the El Niño - Southern Oscillation. J Geophys Res Atmospheres 119:1249-1265 doi:10.1002/2013JD020407

157. Xu J, Smith AK, Wang W, Jiang G, Yuan W, Gao H, et al. (2013) An observational and theoretical study of the longitudinal variation in neutral temperature induced by aurora heating in the lower thermosphere. J Geophys Res Space Phys 118:7410-7425 doi:10.1002/2013JA019144

158. Yamazaki Y, Richmond AD, Yumoto K (2012) Stratospheric warmings and the geomagnetic lunar tide: 1958-2007. J Geophys Res 117: doi:10.1029/2012JA017514

159. Yee JH, Cameron GE, Kusnierkiewicz DY (1999) Overview of TIMED. Proc. SPIE 3756:244 doi:10.1117/12.366378

160. Yi it E, Medvedev AS, Aylward AS, Ridley AA, Harris MJ, Moldwin MB, et al. (2012) Dynamical effects of internal gravity waves in the equinoctial thermosphere. J Atmos Sol Terr Phys 90–91:104-116 doi:10.1016/j.jastp.2011.11.014

161. Yi it E, Medvedev AS, England SL, Immel TJ (2014) Simulated variability of the high-latitude thermosphere induced by small-scale gravity waves during a sudden stratospheric warming. J Geophys Res Space Phys 119:357-365 doi:10.1002/2013JA019283

162. Yizengaw E, Moldwin MB (2009) African Meridian B-Field Education and Research (AMBER) array. Earth Moon Planets 104(1–4):237-246 doi:10.1007/s11038-008-9287-2

163. Yokoyama T, Hysell DL, Otsuka Y, Yamamoto M (2009) Three-dimensional simulation of the coupled Perkins and Es-layer instabilities in the nighttime midlatitude ionosphere. J Geophys Res 114: doi:10.1029/2008JA013789

164. Yumoto K (2011) Update on MAGDAS activities around the globe during ISWI. 2011 IEEE International Conference on Space Science and Communication: 'Towards Exploring the Equatorial Phenomena' IconSpace 2011 Proc art. no. 6015894, 254–257

Chapter 6

Using GIS Data to Build Informed Virtual Geographic Environments (IVGE)

Mehdi Mekni

Department of Math Science and Technology, University of Minnesota, Crookston, USA

ABSTRACT

In this paper, we propose a novel approach to automatically building Informed Virtual Geographic Environments (IVGE) using data provided by Geographic Information Systems (GIS). The obtained IVGE provides

2D and 3D geographic information for visualization and simulation purposes. Conventional VGE approaches are generally built upon a grid-based representation, raising the well-known problems of the lack of accuracy of the localized data and the difficulty to merge data with multiple semantics. On the contrary, our approach uses a topological model and provides an exact representation of GIS data, allowing an accurate geometrical exploitation. Moreover, our model can merge semantic information, even if spatially overlapping. In addition, the proposed IVGE contains spatial information which can be enhanced thanks to a geometric abstraction method. We illustrate this model with an application which automatically extracts the required data from standard GIS files and allows a user to navigate and retrieve information from the computed IVGE.

INTRODUCTION

Modern geography techniques play an irreplaceable role in exploring temporal-spatial patterns and dynamic processes, and understanding the relationships between geographic features, objects, and actors in the real world. Studying these relationships is fundamental and comprehensive geographical analysis often requires the integration of different disciplines at various scales. By using a variety of different techniques, it is possible to produce virtual reality representations, integrated models, simulations, forecasts and evaluations of the geographic environment. Therefore, the Virtual Geographic Environment (VGE) can be a useful tool for understanding the evolutionary process, temporal-spatial patterns, driving mechanisms, and the direction of succession in the real environment. Current VGE techniques, including three-dimensional (3D) techniques, have extended the potential applications of geographic modeling. The ability of VGE to depict past, present, and future geographic environments has been of particular importance [1].

Traditional methods for extracting and expressing spatial information can be modified to better suit development in the field of geographic modeling [2]. Through using geographic analytical modeling and visualization techniques, VGE can be used to perform geographic analysis, simulate geographic phenomena, represent and predict changes in the geographic environment, and evaluate the influence of human activities on the environment [3]. By using VGE to share ideas, people can cooperate and coordinate their work on geographic objects and phenomena, resulting in more advanced methods of designing and transforming the world (Figure 1). VGE will become increasingly important in geography in the future [2].

In this paper, we propose a novel approach to modeling VGE which deals with all these constraints. Our approach provides an exact representation of the geographic environment using vector data and including elevation. This representation is organized as a topological graph, enhanced with data integrating both quantitative (geometry) and qualitative information (zones such as roads and buildings). Moreover, our approach is directed to efficient spatial reasoning like path planning for crowd simulation (Figure 2). Hence, we also address the performance issue when exploiting such environments.

Figure 1: 3D city models of East Berlin.

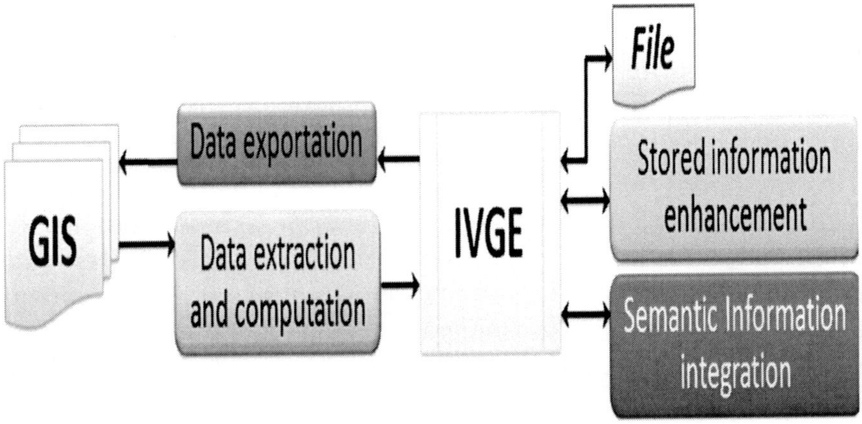

Figure 2: The proposed architecture to generate an IVGE.

We call our model an Informed Virtual Geographic Environment (IVGE) because it is close to standard VGE applications while addressing additional information management issues. IVGE data are directly extracted and computed from standard GIS vector files. It is a one-time automated process based on a Constrained Delaunay Triangulation (CDT) technique [4]. The produced IVGE contains spatial information which can be enhanced thanks to a geometric abstraction method. Finally, the IVGE can either be saved for future use or be exported to a GIS vector file.

The remainder of this chapter finishes with an overview on spatial reasoning and GIS data followed by a discussion of related works on GIS data and the representation of virtual environments. In Section 3, we present our approach to creating an IVGE from GIS data. Section 4 depicts a way to enhance the IVGE information by automatically associating semantics with surfaces based on their boundaries. Next, Section 5 presents two additional information enhancements based on a geometrical abstraction: first, by filtering potential elevation errors related to the input data; second, by qualifying elevation with additional semantics which are associated with specific areas in the environment. Then, Section 6 presents some results obtained by applying our approach to an urban environment. Finally, we provide a summary of this paper and present future work.

SPATIAL REASONING AND GIS DATA

Spatial reasoning is a research area with applications in several domains such as crowd simulation and robotics. Reasoning about space does not only require appropriate computation algorithms, but also an efficient description of the spatial environment. Indeed, such a description must represent the geometrical information which corresponds to geographic features. Moreover, this representation must qualify space by associating semantics to geographic features in order to allow spatial reasoning. A Geographic Information System (GIS) is a system of hardware, software and procedures used to facilitate the management, manipulation, analysis, modeling, representation and display of georeferenced data to solve complex problems regarding planning and management of resources [5]. GISs have emerged in the last decade as an essential tool for urban and resource planning and management [6]. Their capacity to store, retrieve, analyze, model and map large areas with huge volumes of spatial data has led to an extraordinary proliferation of applications [5]. GISs offer two data models to describe a geographic environment: raster and vector data representations [7].

Raster Model

The raster representation of GIS data is a method for the storage, processing and display of spatial data [8]. Each area is divided into rows and columns, which form a regular grid structure [7]. Each cell must be rectangular in shape, but not necessarily square [6]. Each cell within this matrix contains location coordinates as well as an attribute value [8]. The spatial location of each cell is implicitly contained within the ordering of the matrix [9]. Areas containing the same attribute value are recognized as such, however, raster structures cannot identify the boundaries of such areas as polygons [6]. Raster data are an approximation of the real world where spatial data are expressed as a matrix of cells, with spatial position implicit in the ordering of the pixels [9]. With the raster data model, spatial data are not continuous but divided into discrete units [8]. This makes raster data particularly suitable for certain types of spatial operation, for example overlays or area calculations. Raster structures may lead to increased storage in certain situations, since they store each cell in the matrix regardless of whether it is a feature or simply empty space.

Vector Model

The vector format is defined by the vector representation of its geographic data [10]. Vector storage implies the use of vectors (directional lines) to represent a geographic feature [11]. Vector data is characterized by the use of sequential points or vertices to define a linear segment. Each vertex consists of X and Y coordinates [8]. Vector lines are often referred to as arcs and consist of a string of vertices terminated by a node [6]. A node is defined as a vertex that starts or ends an arc segment. Point features are defined by one coordinate pair, a vertex. Polygonal features are defined by a set of closed coordinate pairs [7]. In vector representation, the storage of the vertices for each feature is important, as well as the connectivity between features, e.g. the sharing of common vertices where features connect [9].

GIS offer two ways to describe an environment: grid and vector representations. Grids have several drawbacks such as the difficulty to balance accuracy and memory use, making them unusable for the precise exploitation of large scale environments. Vector layers are scalable, but it is difficult to manipulate the stored data in order to simply retrieve pieces of information or to merge data with different semantics. Moreover, while GIS data are stored in a quantitative way which suits to exact calculations, spatial reasoning often needs to manipulate qualitative information. For example, when considering a slope in a landscape, it is simpler and faster to qualify it by an attribute that takes the values light or steep rather than to directly use its angle value with respect to the horizontal plane. This process of qualification, which associates semantics with quantitative intervals of values, greatly simplifies the description and validation of reasoning mechanisms. Moreover, the slope example illustrates another requirement of spatial reasoning which is poorly addressed in the literature: providing a way to handle terrain's elevations. Indeed, a real environment is rarely flat and ignoring this information would greatly decrease the quality of spatial reasoning.

RELATED WORK

GIS data are usually available in either raster or vector formats. The raster format subdivides the space into regular square cells, called boxes, associated with space related attributes. This approach generally presents average quantitative data whose precision depends on the scale of the representation. In contrast, the vector format exactly describes geographic information without constraining geometric shapes, and generally associates one qualitative data with each shape. Such data are usually exploited by a VGE in two ways [3]: the grid method and the exact geometric subdivision. The grid method is the direct mapping of the raster format [12], but it can also be applied to the vector format (Figure 3(a)). This discrete representation can be used to merge multiple semantic data [13], the locations where to store these data being predefined by the grid cells. The main drawback of the grid method is related to a loss in location accuracy [1], making it difficult to accurately position any information which is not aligned with the subdivision. Another drawback arises when trying to precisely represent large environments using a grid: the number of cells tends to increase dramatically, which makes the environment exploitation very costly [14]. The grid-based method is mainly used for animation purposes [15] and large crowd simulations [16] because of the fast data access it provides.

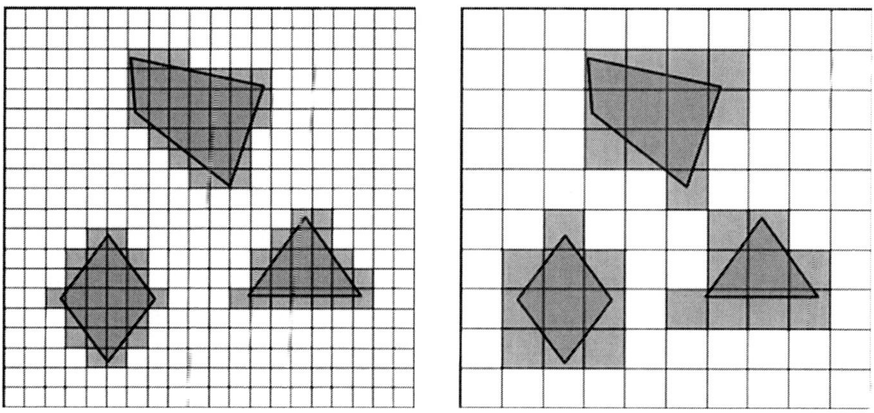

(a)

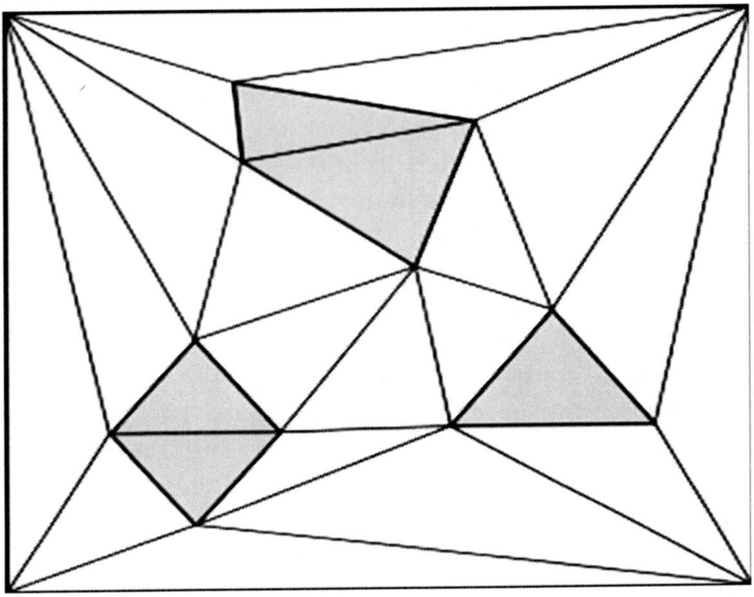

(b)

Figure 3: Two common cell decomposition techniques: (a) approximate decomposition by grids considering two resolutions. White boxes are free, grey are obstacles; (b) exact decomposition using CDT.

The second method, called exact geometric subdivision, consists in subdividing the environment in convex cells using the vector format as an input. The convex cells can be generated by several algorithms, among which the most popular is the Constrained Delaunay Triangulation (CDT) [17]. This produces triangles while keeping the original geometry segments which are named constraints (Figure 3(b)). The first advantage of the exact subdivision method is to preserve the input geometry, allowing accurately manipulating and visualizing the environment at different scales. Another advantage of this approach is that the number of produced cells only depends on the complexity of the input shapes, but not on the environment's size and scale as it is the case with the grid method [18]. The main drawback of this approach is the difficulty to merge multiple semantic data for overlapping shapes. Moreover, this method is generally used to represent planar environments because the CDT can only handle 2D geometries [19].

This method tends to be used for crowd microscopic simulations [2] where motion accuracy is essential.

Both VGE representations, approximate and exact, can be enhanced by an abstraction process. The first goal of an abstraction is to improve the performance of the algorithms based on the environment description, such as path planning, by reducing the number of elements used to describe the environment. The usual abstraction model for grids is mainly geometric: the quadtree groups four boxes of the same kind to create a higher level cell [20]. When considering the exact decomposition, an abstraction is generally based on topological properties rather than on purely geometric ones. Indeed, the exact cell subdivision generates connected triangles which can be manipulated as the nodes of a topological graph. This graph can then be abstracted by grouping the nodes, producing a new graph with fewer nodes [21]. For example, Figure 4 shows an abstraction which is only based on the nodes' number of connections c: isolated (c = 0), deadend (c = 1), corridor (c = 2), and crossroad (c = 3). A topological graph can be used for spatial reasoning, like path planning, thanks to traversal algorithms. These algorithms benefit of the abstraction by traversing first the more abstracted graph, and then by refining the computation in the sub-graphs until reaching the graph of the spatial subdivision [22]. This exploitation points out a new need for an abstracted graph which is less addressed in literature: it must contain the minimal information necessary to make a decision. For example, if the width of a path is relevant for a path planning algorithm, this information must be accessible in all the abstracted graphs; if not, the evaluation would be greatly distorted compared to a un-abstracted graph.

Two kinds of information can be stored in the description of an IVGE. Quantitative data are stored as numerical values which are generally used to depict geometric properties (path's width of 2 meters) or statistical values (density of 2.5 persons per square meter). Qualitative data are introduced as identifiers which can be a reference to an external database or a word with an arbitrary semantic, called a label.

Such labels can be used to qualify an area (a road or a building) or to interpret a quantitative value (a narrow passage or a crowded place). An advantage of interpreting quantitative data is to reduce a potentially infinite set of inputs to a discrete set of values, which is particularly useful to condense information in successive abstraction levels to be used for reasoning purposes.

The approach we propose is based on an exact representation whose precision allows realistic applications like crowds' micro-simulations. We will briefly describe how to obtain such a representation since it has already been presented in [16]. We will then show how the obtained topological graph can be improved in two ways. First, by propagating input qualitative information from the arcs of the graph to the nodes, this allows for example to deduce the internal parts of the buildings or of the roads in addition to their outline. Second, we propose a novel approach of information extrapolation using a onetime spatial reasoning process based on a geometric abstraction. This second technique can be used to fix input elevation errors, as well as to create new qualitative data relative to elevation variations. These data are stored as additional semantics bound to the graph nodes, which can subsequently be used for spatial reasoning.

AUTOMATED GENERATION OF IVGE

We propose an automated approach to compute the data directly from vector format GIS data. This approach is based on three stages which are detailed in this section (Figure 5): Input data selection, spatial decomposition, and maps unification.

The first step of our approach, the inputting data selection is the only one requiring human intervention. The various vector data selected which will be used to build the IVGE. The only restrictions concerning these data are that they need to respect the same scale and be equally geo-referenced.

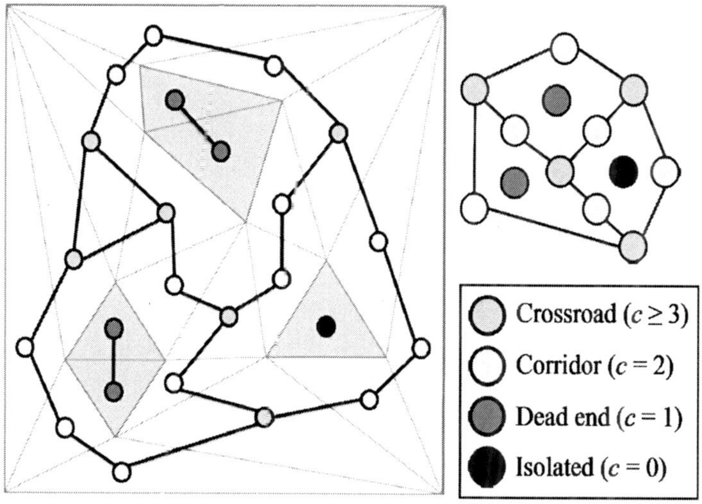

Figure 4: Topologic abstraction of virtual environments.

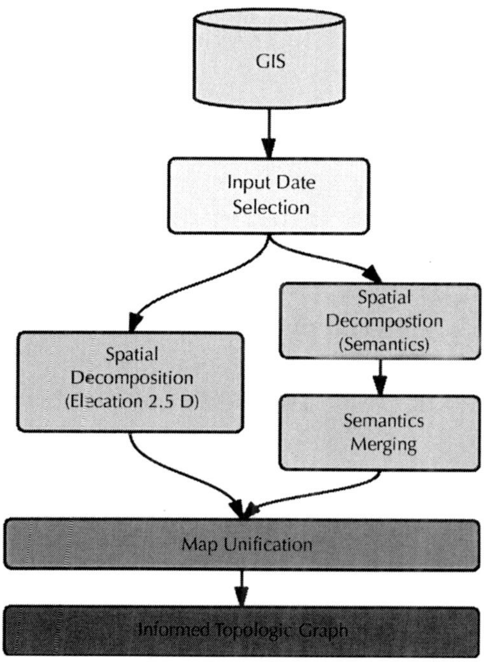

Figure 5: A flow chart illustrating the automated generation of IVGE.

The input data can be organized in two categories. First, elevation layers contain geographical marks indicating absolute terrain elevations. Since we consider the creation of 2.5D, a given coordinate cannot have two different elevations, prohibiting the representation of tunnels for example. Moreover, several elevation layers can be specified, the model being able to merge them automatically. Second, semantic layers are used to qualify various features of the geographic space. As shown in Figure 4, each layer indicates the geographic boundaries of a set of features having identical semantics, such as roads and buildings. The boundaries of the features can overlap between two layers, our model is able to merge this information.

The second step of our method, spatial decomposition, consists of obtaining an exact spatial decomposition of the input data in cells. This process uses a Delaunay triangulation and is entirely automatic. It and can be divided into two parts in relation to the previous phase. First, an elevation map is computed, corresponding to the triangulation of the elevation layers. All the elevation points of the layers are injected in a 2D triangulation, the elevation is considered as an additional datum. This process produces an environment subdivision composed of connected triangles. Such a subdivision provides information about coplanar areas: the elevation of any point inside the environment can be deduced thanks to the elevation of the three vertices of the corresponding triangle. Second, a merged semantics map is computed, which corresponds to a Constrained Delaunay Triangulation (CDT) of the semantic layers. Indeed, each segment of a semantic layer is injected as a constraint which keeps track of the original semantic data thanks to an additional datum. Consequently, the resulting map is a merging all input semantics: each constraint represents as many semantics as the number of input layers containing it.

In the third step, the two maps previously obtained are merged. This phase corresponds to the mapping of the 2D merged semantic map on the 2.5D elevation map in order to obtain the final 2.5D elevated merged semantics map. First, a preprocessing is carried out on the merged semantics map in order to preserve the elevation precision inside the unified map. Indeed, all the points of the elevation map are injected in the merged semantics triangulation, and create new triangles. Then, a second process elevates the merged semantics map. The elevation of each merged semantics point P is computed by retrieving the triangle T of the elevation map whose 2D projection contains P. Once T is obtained, the elevation is simply computed by projecting P on the plane defined by T using the Z axis. When P is outside the convex hull of the elevation map, no triangle can be found and the elevation cannot be directly deduced. In this case, we use the average height of the points of the convex hull which are visible from P.

SEMANTIC INFORMATION MANAGEMENT

The obtained results of the importation technique emphasize qualified areas by defining the semantics of their boundaries. But these informed boundaries are difficult to exploit when dealing with the semantics associated with a position. For example, it is difficult to check if a position is inside a building. This is why we propose to enhance the information provided by spreading the boundaries' semantics to the cells. Three related processes are necessary, and explained in the following subsections: graph analysis, potential conflicts resolution, and semantics assignment.

Graph Analysis

The graph analysis is a traversal algorithm which explores the environmental graph while qualifying the cells towards a given semantic. This algorithm is applied to the entire graph one time for each semantic to propagate. While exploring the graph, the algorithm collects three kinds of cells which are stored in three container structures for future use: Inside, Outside and Conflict. Cells are within an area delimited by borders associated to the propagated semantic. Outside cells are outside any area defining the propagated semantic. Conflict (C) cells are both qualified inside and outside by the algorithm. Non Conflict (NC) cells are either qualified inside or outside by the algorithm. Three parameters influence the traversal: 1) the semantic sem to propagate. 2) a set of starting cells, indicating where to start the exploration of the graph; a set is provided instead of a single cell in order to be able to manage disconnected graphs. 3) a boolean value startin indicating whether the semantic must be assigned to the starting cells or not. The recursive algorithm is detailed in Figure 6.

Resolution of Conflicts

After each graph traversal, we must deal with the cells that are potentially in conflict. Indeed, these cells must be assigned to either the Inside or the Outside container in order for the system to continue with the next phase. Cells are in conflict when the shapes of two input features with the same semantic share a segment. Two alternative methods are proposed: 1) a fast assignment where the conflicting cells are arbitrarily transferred to one of the target containers, and 2) a deductive assignment where an algorithm selects the best option based on geometric considerations. The arbitrary assignment is used when the internal details of a shape are not relevant for the target application. The deductive assignment is used when the internal details of a shape are relevant. Both methods are carried out in two steps: 1) a local conflicting graph extraction which is the same for both methods, and 2) a decision step which is specific to each method. The local conflicting graph extraction collects all the cells surrounding a conflicting cell, but only if they are reachable through a border which is not marked by the propagated semantic. Each orange zone in Figure 6 shows an extracted local conflicting graph. Every time a cell is discovered, it is transferred from the global container Conflict to a local container. Then, the algorithm recursively explores the neighbors which are reachable through a border which is not marked by the propagated semantic, transferring them to the local container. At the end, the algorithm obtains a set of local conflict containers, corresponding to the amount of local graphs considered in conflict.

The decision part of the arbitrary assignment only consists in transferring the local conflicting cells to one of the Inside or the Outside containers. The decision part of the deducing algorithm is based on geometric considerations. If the local conflicting zone is mainly surrounded by Outside cells, then the conflict is resolved as Inside, and vice versa. The conflict management algorithm is described in Figure 7.

Semantics Assignment

The last step of the semantic propagation consists in assigning the final semantics to the cells. The process is very simple: each propagated semantic is assigned to all the corresponding Inside cells. One can notice that some cells may have multiple semantics when they are present in more than one Inside container. Additionally, it is possible to keep track of the Outside cells by assigning them a negative semantic, as for example in order to know what is road in the environment and what is not.

Algorithm 2 Graph Analysis

PARAMETERS: ***crt_sem*** the propagated semantic label; ***start_cells*** the starting cells; ***start_inside*** defining if starting cells are inside or outside.

REQUIRE: ***inside_cells*** the container of inside cells; ***outside_cells*** the container of outside cells.

```
FOR ALL cell ∈ start_cells DO
    IF start_inside THEN
        TRAVERSE (cell, inside_cells, outside_cells)
    ELSE
        TRAVERSE (cell, outside_cells, inside_cells)
    END IF
END FOR
```

Figure 6: Graph analysis algorithm.

Algorithm 3 Traverse

PARAMETERS: ***crt_cell*** the current cell; ***crt_cont*** the current cell container; ***oth_cont*** the other container.

REQUIRE: ***crt_sem*** the propagated semantic label; ***conflict_cells*** the container of conflicting cells.

```
IF crt_cell ∉ crt_cont ∧ crt_cell ∉ conflict_cells THEN
    IF crt_cell ∈ oth_cont THEN
        oth_cont ← oth_cont \ crt_cell
        conflict_cells ← conflict_cells ∪ {crt_cell}
    ELSE
        crt_cont ← crt_cont ∪ {crt_cell}
        FOR i = 0 to NEIGHBOURS NUMBER (crt_cell) DO
            nxt_border ← BORDER (crt_cell, i)
            nxt_cell ← NEIGHBOURS (crt_cell, i)
            IF crt_sem ∈ SEMANTICS OF (nxt_border) THEN
                TRAVERSE (nxt_cell, oth_cont, crt_cont)
            ELSE
                TRAVERSE (nxt_cell, crt_cont, oth_cont)
            END IF
        END FOR
    END IF
END IF
```

Figure 7: Semantic propagation and conflict management algorithm.

Finally, an optional process can be performed to remove the borders' semantics of some detected conflicting cells. Indeed, such borders may distort some spatial reasoning algorithms. For example, when considering road borders as obstacles to plan a path, a simulated vehicle would not be able to go through some passageways. After resolution, the semantic of the problematic borders is removed, making them crossable. These problematic borders are the ones which are marked with a propagated semantic and which connect two Inside cells. One can note that only the cells previously detected as conflicting need to be tested.

GEOMETRIC ABSTRACTION OF IVGE

Spatial decomposition subdivides the environment into convex cells. Such cells encapsulate various quantitative geometric data which are suitable for accurate computations. Since geographic environments are seldom flat, it is important to consider the terrain's elevation, which is quantitative geometric data. Moreover, while elevation data are stored in a quantitative way which suits exact calculations, spatial reasoning often needs to manipulate qualitative information. Indeed, when considering a slope, it is obviously simpler and faster to qualify it using an attribute such as light and steep rather than using numerical values. However, when dealing with large scale geographic environments, handling the terrain's elevation, including its light variations, may be a complex task. To this end, we propose an abstraction process that uses geometric data to extract the average terrain's elevation information from spatial areas. The objectives of this Geometric Abstraction are threefold [23]. First, it aims to reduce the amount of data used to describe the environment. Second, it helps detect anomalies, deviations, and aberrations in elevation data. Third, the geometric abstraction enhances the environmental description by integrating qualitative information characterizing the terrain's elevations. In this section, we first present the algorithm which computes the geometric abstraction. Then, we describe two processes which use the geometric abstraction: Filtering elevation anomalies and Extracting elevation semantics.

Geometric Abstraction Algorithm

The geometric abstraction process gathers cells in groups according to a geometric criterion: we chose the coplanarity of connex cells in order to obtain uniform elevation areas. The algorithm takes advantage of the structure obtained thanks to the IVGE generation process.

The aim of this algorithm is to group cells which verify a geometric criterion in order to build groups of cells. A cell corresponds to a node in the topological graph. A node represents a triangle generated by the CDT spatial decomposition technique. A cell is characterized by its boundaries, neighboring cells, and surface as well as its normal vector which is a vector perpendicular to its plan Figure 8(a).

A group is a container of adjacent cells. The grouping strategy of this algorithm is based on a coplanarity criterion which is assessed by computing the difference between the normal vectors of two neighboring cells or groups of cells (Figure 8(a)). Since a group is basically composed of adjacent cells it is obvious to characterize a group by its boundaries, its neighboring groups, its surface, as well as its normal vector. However, the normal vector of a group must rely on an interpretation of the normal vectors of its composing cells. In order to compute the normal vector of a group, we adopt the area weight normal vector [24] which takes into account the unit normal vectors of its composing cells as well as their respective surfaces. The area-weight normal vector of a group $\vec{N}G$ is computed using:

$$\vec{NG} = \sum_{c \in G} \left(S_c \cdot \vec{N_c} \right) \Big/ \sum_{c \in G} S_c \quad (1)$$

where S_c denotes the surface of a cell c and N_c its unit normal vector. Hence, thanks to the area-weight normal vector we are able to compute a normal vector for a group based on the characteristics of its composing cells. The geometric abstraction algorithm uses two input parameters: 1) a set of starting cells which act as access points to the graph structure, and 2) a gradient parameter which corresponds to the maximal allowed difference between cells' inclinations. Indeed, two adjacent cells are considered coplanar and grouped together when the angle between their normal vectors (α in Figure 8(b)) is lesser than gradient.

The analysis of the resulting groups helps to identify anomalies in elevation data. Such anomalies need to be fixed in order to build a realistic virtual geographic environment. Furthermore, the average terrain's elevation which characterizes each group is quantitative data described using area-weighted normal vectors. Such quantitative data are too precise to be used by qualitative spatial reasoning. For example, when considering a slope in a landscape, it is simpler and faster to qualify it using a simple attribute that takes the values light and steep rather than to express it using the angle value with respect to the horizontal plane. Hence, a qualification process which aims to associate semantics with quantitative intervals of values characterizing a group's terrain inclinations would greatly simplify spatial reasoning mechanisms. In addition, in order to fix anomalies in elevation data and to qualify the groups' terrain inclinations we propose to apply a technique to improve the geometric abstraction of the VGE. The geometric abstraction allows improving IVGE by filtering the elevation anomalies, qualifying the terrain's elevation using semantics and integrating such semantics in the description of the geographic environment.

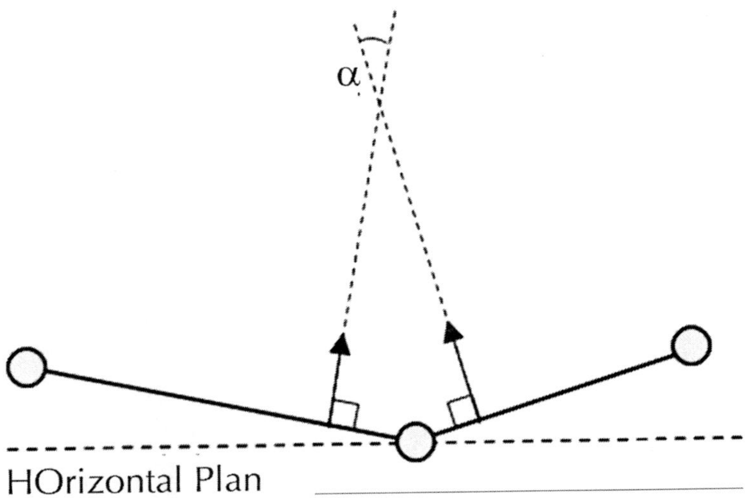

(a)

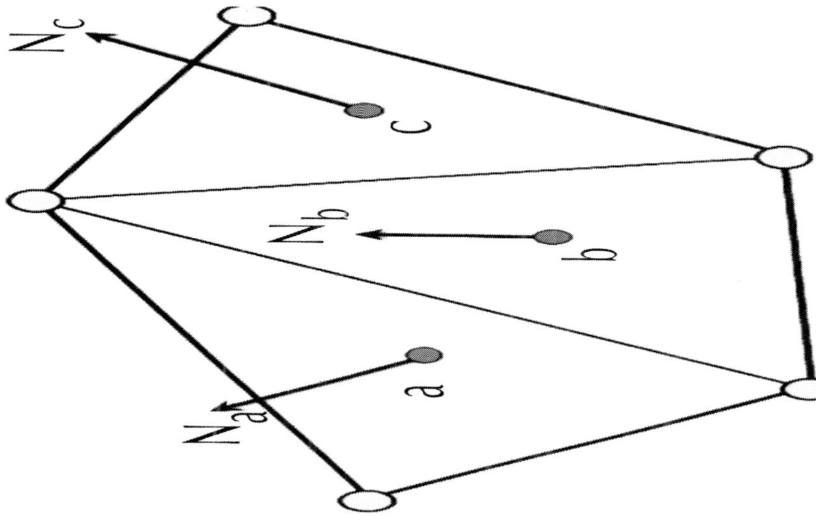

(b)

Figure 8: (a) Illustration of two coplanar cells; (b) Unit normal vectors.

A detailed description of the geometric abstraction algorithm as well as GIS data filtering and elevation qualification processes are provided in [23]. The geometric abstraction is built using a graph traversal algorithm. It groups cells based on their area-weighted normal vectors (Figure 8(b)). The objectives of the geometric abstraction are threefold. First, it qualifies the terrain's elevation of geographic environments to simplify spatial reasoning mechanisms. Second, it helps identifying and fixing elevation anomalies in initial GIS data. Third, it enriches the description of geographic environments by integrating elevation semantics.

Filtering Elevation Anomalies

The analysis of the geometric abstraction may reveal some isolated groups which are totally surrounded by a single coherent group. These groups are characterized by a large difference between their respective area-weighted normal vectors. Such isolated groups are often characterized by their small surfaces and can be considered anomalies, deviations, or aberrations in the initial elevation data. The geometric abstraction process helps to identify and allows automatically filtering such anomalies thanks to a two phase process.

First, isolated groups are identified (Figure 9(a)). The identification of isolated groups is based on two key parameters: 1) the ratio between the surface of surrounded and surrounding groups, and 2) the difference between the area-weighted normal vectors of the surrounded and surrounding groups.

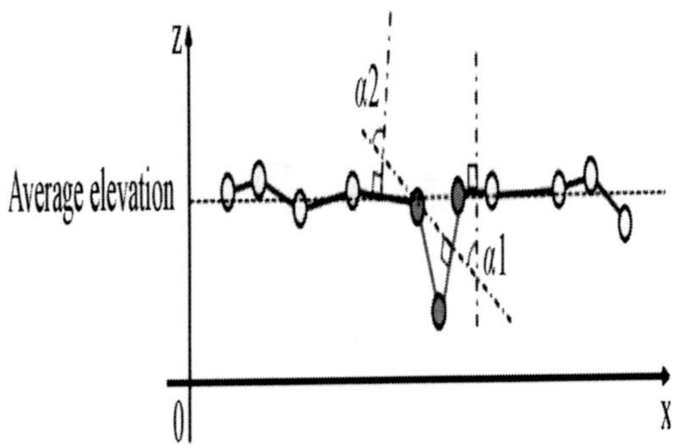

(a)

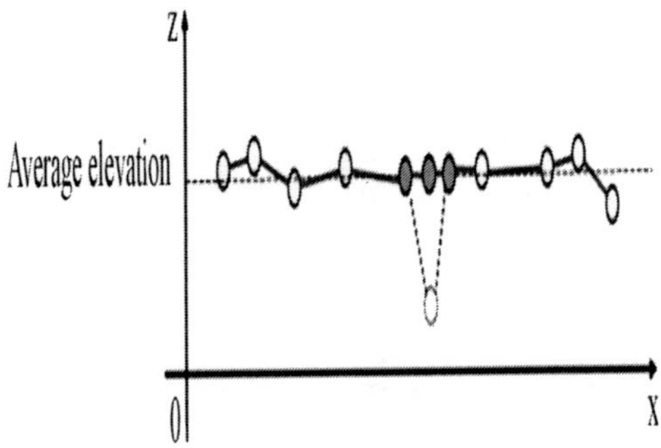

(b)

Figure 9: Profile section of anomalous isolated groups (red color) adjusted to the average elevation of the surrounding ones (yellow color).

Second, these isolated groups are elevated at the average level of elevation of the surrounding ones (Figure 9(b)). In fact, the lowest and the highest elevations of the isolated group are computed. Then, the elevations of all the vertices of isolated grp are updated using the average between the lowest and highest elevation. As a consequence, we obtain more coherent groups in which anomalies of elevation data are corrected.

Extracting Elevation Semantics

The geometric abstraction algorithm computes quantitative geometric data describing the terrain's inclination. Such data are stored as numerical values which allow to accurately characterize the terrains' elevations.

However, handling and exploiting quantitative data is a complex task as the volume of values may be too large and as a consequence difficult to transcribe and analyze. Therefore, we propose to interpret the quantitative data of the terrain's inclination by qualifying the areas' elevations. Semantic labels, called semantics elevation, are associated to quantitative intervals of values that represent the terrain's elevation. In order to obtain elevation semantics we propose a two steps process taking advantage of the geometric abstraction: 1) a discretization of the angle between the weighted normal vector Ng of a group g and the horizontal plane, 2) an assignment of semantic information to each discrete value which qualifies it. The discretization process can be done in two ways: a customized and automated approach.

The customized approach qualifies the terrain's elevation and requires that the user provides a complete specification of the discretization. Indeed, the user needs to specify a list of angle intervals as well as their associated semantic attributes. The algorithm iterates over the groups obtained by the geometric abstraction. For each group grp, it retrieves the terrain inclination value I. Then, this process checks the interval bounds and determines in which one falls the inclination value I. Finally, the customized discretization extracts the semantic elevation from the selected elevation interval and assigns it to the group grp. For example, let us consider the following inclination interval and the associated semantic elevations: f([10; 20]; gentle slope); f([20; 25]; steep slope). Such a customized specification associates the semantic elevation "gentle slope" to inclination values included in the interval [10; 20] and the semantic elevation "steep slope" to inclination values included in the interval [20; 25]. The automated approach only relies on a list of semantic elevations representing the elevation qualifications. Let N be the number of elements of this list, and T be the total number of groups obtained by the geometric abstraction algorithm. First, the automated discretization orders groups based on their terrain inclination. Then, it iterates over these ordered groups and uniformly associates a new semantic elevation from the semantics set of each T = N processed groups. For example, let us consider the following semantic elevations: gentle; medium and steep. Besides, let us consider an ordered set of groups S which may be composed of six groups of cells.

Let us compare these two discretization approaches. On the one hand, the customized discretization process allows one to freely specify the qualification of the terrain's elevations. However, qualifications resulting from such a flexible approach deeply rely on the correctness of the interval bounds' values. Therefore, the customized discretization method requires having a good knowledge of the terrain characteristics in order to guarantee a valid specification of inclination intervals. On the other hand, the automated discretization process is also able to qualify groups' elevations without the need to specify elevation intervals' bounds. Such a qualification usually produces a visually uniformed semantic assignment. This method also guarantees that all the specified semantic attributes will be assigned to the groups without a prior knowledge of the environment characteristics.

Enhancing the Geometric Abstraction

The geometric abstraction algorithm produces groups that are built on the basis of their terrain's inclination characteristics. Thanks to the extraction of elevation semantics, the terrain's inclination is qualified using semantic attributes and associated with groups and with their cells. Because of the game of the classification intervals, adjacent groups with different area-weighted normal vectors may obtain the same elevation semantic. In order to improve the results provided by the geometric abstraction, we propose a process that merges adjacent groups which share the same semantic elevation. This process starts by iterating over groups. Then, every time it finds a set of groups sharing an identical semantic elevation, it creates a new group. Next, cells composing the adjacent groups are registered as members of the new group. Finally, the area-weighted normal vector is computed for the new group. Hence, this process guarantees that every group is only surrounded by groups which have different semantic elevations (Figure 10). In this section, we proposed a geometric abstraction process as well as three heuristics which take advantage of this process. The geometric abstraction is built using a graph traversal algorithm. It groups cells based on their areaweighted normal vectors. The objectives of the geometric abstraction are threefold. First, it qualifies the terrain's elevation of geographic environments to simplify spatial reasoning mechanisms. Second, it helps identify and fix elevation anomalies in initial GIS data. Third, it enriches the description of geographic environments by integrating elevation semantics.

RESULTS

The proposed environment extraction method is used to create an accurate IVGE and provides the advantages of standard GIS visualization techniques including the semantic merging of grids along with the accuracy of vector data layers. Thanks to the automatic extraction method that we propose, our system handles the IVGE construction directly from a specified set of vector format GIS files. The performance of the extraction process is very good and able to process an area such as the center part of Quebec-City, with one elevation map and five semantic layers, in less than five seconds on a standard computer (Intel Core 2 Duo processor 2.13 Ghz, 1 G RAM). The obtained unified map approximately contains 122,000 triangles covering an area of 30 km^2. Besides, the necessary time to obtain the triangle corresponding to a given coordinate is negligible (less than 10^{-4} seconds). Moreover, the geometric abstraction produces approximately 73,000 groups in 2.8 seconds. Finally, the custom and automated discretization processes are performed respectively in 1.8 and 1.2 seconds using eight semantic elevation labels. The IVGE application provides two visualization modes for the computed data. First, a 3D view (Figure 10) allows to freely navigating the virtual environment. We propose an optional mode for this view where the camera is constrained at a given height above the ground, allowing the elevation variations to be followed when navigating. Second, we propose an upper view with orthogonal projection to represent the GIS data as a standard map. In this view the user can scroll and zoom the map (Figure 10), and can accurately view any portion of the environment at any scale. Additionally, one can select a position in the environment in order to retrieve the corresponding data (2 in Figure 10), such as the underlying triangle geometry, the corresponding height, and the associated semantics, including semantic elevation.

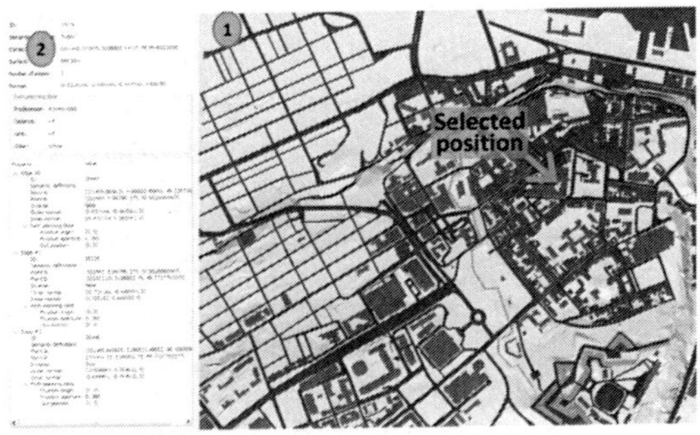

(a)

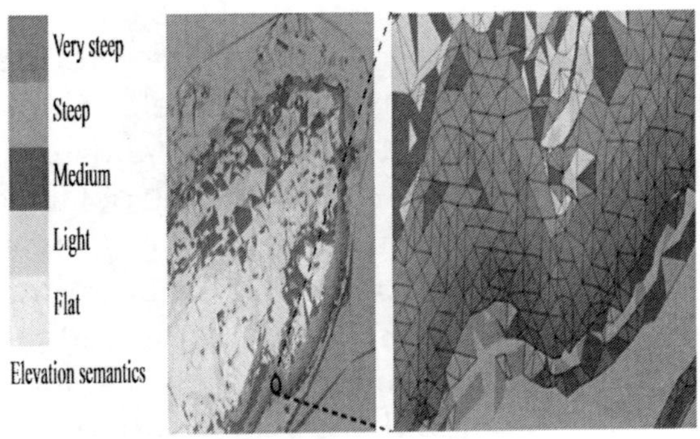

(b)

Figure 10: (a) 2D map visualization of the IVGE: (1) Unified map, (2) Accurate information about the selected position; (b) Results of the geometric abstraction process.

Our approach goes beyond classical grid-based visualization by combining the semantic information and the vector-based representation's accuracy. Indeed, our topological method combines the advantages of grids and vector layers, and avoids their respective drawbacks. Moreover, this data extraction method is fully automated, being able to directly process GIS vector data. In addition, the propagation of the borders' semantics enhances the information provided by the IVGE. As a result, both boundaries and areas are semantically qualified (roads and buildings). Besides, the geometric abstraction algorithm enriches the IVGE description by qualifying the terrain inclination characteristics of cells using semantic elevation. Finally, we have shown the suitability of this method for GIS visualization thanks to an IVGE application which allows two visualization modes: 3D for immersion purpose, and 2D to facilitate geographic data analysis.

CONCLUSIONS

All of the above-mentioned characteristics allow us to anticipate several applications of this work, mainly thanks to the exploitation of the topological graph. Firstwe are currently working on the application of the IVGE for crowd simulation. Indeed, this informed environment is particularly well adapted to such a domain, since it allows efficient path planning and navigation algorithms and provides useful spatial data which are needed for agents' behaviors. However, we plan to further improve the environment description by using topological graph abstractions. This will allow us to reduce the complexity of the graph exploration algorithms, and to deduce additional properties such as the reachable areas for a human being with respect to variations of land slopes. Second, we want to append new information to the environment description in order to represent moving and stationary situated elements. These elements could be humans, vehicles, or even street signs, depending on the objectives of the virtual reality application. Our topological graph which describes the geographic environment can be extended by integrating conceptual nodes representing such elements. Thanks to the geometrical accuracy of our approach, it will be relatively easy to add this information at any position.

ACKNOWLEDGEMENTS

This research was supported in part by the Research Stimulus Dollars Award provided by the University of Minnesota Crookston. The author would like to thank the reviewers for their valuable comments.

REFERENCES

1. R. Andersen, J. L. Berrou and A. Gerodimos, "On Some Limitations of Grid-Based (CA) Pedestrian Simulation Models," In VRlab, EPFL, 2005.
2. M. Kallmann, H. Bieri and D. Thalmann, "Fully Dynamic Constrained Delaunay Triangulations," Geometric Modelling for Scientific Visualization, 2003.
3. F. Lamarche and S. Donikian, "Crowds of Virtual Humans: A New Approach for Real Time Navigation in Complex and Structured Environments," Computer Graphics Forum, Eurographics'04, 2004.
4. J. Gong and L. Hui, "Virtual Geographical Environments: Concept, Design, and Applications," 1999.
5. D. Demyen and M. Buro, "Efficient Triangulation-Based Pathfinding," 2006.
6. M. F. Goodchild, "GIS and Disasters: Planning for Catastrophe," Computers, Environment and Urban Systems, Vol. 30, No. 3, 2006, pp. 227-229.http://dx.doi.org/10.1016/j.compenvurbsys.2005.10.004
7. N. Chrisman, "Exploring Geographical Information Systems," 2nd Edition, John Wiley and Sons Inc., 2001.
8. D. Arctur and M. Zeiler, "Designing Geodatabases: Case Studies in GIS Data Modeling," ESRI Press, 2004.
9. P. Longley, M. Goodchild, M. David and D. Rhind, "Geographic Information Systems and Science," Wiley and Son. Inc., ESRI Press, 2002.
10. S. Fortheringham and P. Rogerson, "Spatial Anaysis and GIS's," Taylor & Francis, LTD, 2002.

11. L. Zhou, G. Lu, Y. Sheng, H. Xu and H. Wang, "A 3D GIS Spatial Data Model Based On Cell Complex," The International Archives of the Photogrammetry, Remote Sensing and Spatial Information Sciences, Vol. 37, 2008, pp. 905-908.
12. M. Mekni, N. Sahli, B. Moulin and H. Haddad, "Using Multi-Agent Geo-Simulation Techniques for the Detection of Risky Areas for Trains," Simulation, Vol. 86, No. 12, 2010, pp. 763-775. http://dx.doi.org/10.1177/0037549709359354
13. J. Zhu, J. Gong, H. Lin, W. Li, J. Zhang and X. Wu, "Spatial Analysis Services in Virtual Geographic Environment Based on Grid Technologies," MIPPR 2005: Geospatial Information, Data Mining, and Applications, Vol. 6045, No. 1, 2005, pp. 604-615.
14. M. Mekni and B. Moulin, "Holonic Modelling of Large Scale Geographic Environments," 2007.
15. F. Tecchia, C. Loscos and Y. Chrysanthou, "Visualizing Crowds in Real-Time," Computer Graphics Forum, Vol. 21, No. 4, 2002, pp. 753-765. http://dx.doi.org/10.1111/1467-8659.00633
16. M. Mekni, "Crowd Simulation Using Informed Virtual Geographic Environments (IVGE)," 2012.
17. A. Botea, M. Muller and J. Schaeffer, "Near Optimal Hierarchical Path-Finding," Journal of Game Development, Vol. 1, 2004, pp. 7-28.
18. L. G. Da and S. R. Musse, "Real-Time Generation of Populated Virtual Cities," 2006.
19. W. Shao and D. Terzopoulos, "Environmental Modeling for Autonomous Virtual Pedestrians," Digital Human Modeling for Design and Engineering Symposium, 2005.
20. S. Paris, S. Donikian and N. Bonvalet, "Environmental Abstraction and Path Planning Techniques for Realistic Crowd Simulation," Computer Animation and Virtual Worlds, Vol. 17, No. 3-4, 2006, pp. 325-335. http://dx.doi.org/10.1002/cav.136
21. D. Brown, "Agent-Based Models," The Earths Changing Land: An Encyclopaedia of Land-Use and Land-Cover Change, Greenwood Publishing Group, Westport, 2006, pp. 7-13.
22. M. Mekni, "Abstraction of Informed Virtual Geographic Environments," Geo-Spatial Information Science, Vol. 15, No. 1, 2012, pp. 27-36.http://dx.doi.org/10.1080/10095020.2012.708150

23. S.-G. Chen and J.-Y. Wu, “A Geometric Interpretation of Weighted Normal Vectors and Its Improvements,” 2005.
24. GDAL: Geospatial Data Abstraction Library.

Citations

CHAPTER 1

Humberto Barbosa, Carolien Tote, Lakshmi Kumar, and Yazidhi, Bamutaze (2013). Harnessing Earth Observation and Satellite Information for Monitoring Desertification, Drought and Agricultural Activities in Developing Countries, Environmental Change and Sustainability, Dr. Steven Silvern (Ed.), ISBN: 978-953-51-1094-1, InTech, DOI: 10.5772/55499.

CHAPTER 2

Rodrigo Nobrega, Colin Brooks, Charles O'Hara and Bethany Stich (2012). Multi-Scale GIS Data-Driven Method for Early Assessment of Wetlands Impacted by Transportation Corridors, Application of Geographic Information Systems, Dr. Bhuiyan Monwar Alam (Ed.), ISBN: 978-953-51-0824-5, InTech, DOI: 10.5772/50349.

CHAPTER 3

Jung-Sup Um (2012), Feng-Shui Theory and Practice Investigated by Spatial Regression Modeling, Application of Geographic Information Systems, Dr. Bhuiyan Monwar Alam (Ed.), ISBN: 978-953-51-0824-5, InTech, DOI: 10.5772/47925.

CHAPTER 4

David Sanz, Santiago Castaño and Juan José Gómez-Alday (2012). GIS Applied to the Hydrogeologic Characterization – Examples for Mancha Oriental Aquifer (SE Spain), Application of Geographic Information Systems, Dr. Bhuiyan Monwar Alam (Ed.), ISBN: 978-953-51-0824-5, InTech, DOI: 10.5772/47967.

CHAPTER 5

Jens Oberheide, Kazuo Shiokawa, Subramanian Gurubaran, William E Ward, Hitoshi Fujiwara, Michael J Kosch, Jonathan J Makela, and Hisao Takahashi, The Geospace Response To Variable Inputs from the Lower Atmosphere: A Review of the Progress Made by Task Group 4 of CAWSES-II, doi:10.1186/s40645-014-0031-4.

CHAPTER 6

M. Mekni, "Using GIS Data to Build Informed Virtual Geographic Environments (IVGE)," Journal of Geographic Information System, Vol. 5 No. 6, 2013, pp. 548-558. doi: 10.4236/jgis.2013.56052.

Index

A

C

E